Contents

1 NUMBER SETS
    1.1   Set Identities **1**
    1.2   Sets of Numbers **5**
    1.3   Basic Identities **7**
    1.4   Complex Numbers **8**

2 ALGEBRA
            Factoring Formulas **12**
    2.2   Product Formulas **13**
    2.3   Powers **14**
    2.4   Roots **15**
            Logarithms **16**
    2.6   Equations **18**
    2.7   Inequalities **19**
    2.8   Compound Interest Formulas **22**

3 GEOMETRY
            Right Triangle **24**
    3.2   Isosceles Triangle **27**
    3.3   Equilateral Triangle **28**
    3.4   Scalene Triangle **29**
    3.5   Square **33**
    3.6   Rectangle **34**
    3.7   Parallelogram **35**
    3.8   Rhombus **36**
    3.9   Trapezoid **37**
    3.10 Isosceles Trapezoid **38**
    3.11 Isosceles Trapezoid with Inscribed Circle **40**
    3.12 Trapezoid with Inscribed Circle **41**

3.13	Kite 42	
	Cyclic Quadrilateral 43	
3.15	Tangential Quadrilateral 45	
3.16	General Quadrilateral 46	
3.17	Regular Hexagon 47	
3.18	Regular Polygon 48	
3.19	Circle 50	
3.20	Sector of a Circle 53	
3.21	Segment of a Circle 54	
3.22	Cube 55	
3.23	Rectangular Parallelepiped 56	
3.24	Prism 57	
3.25	Regular Tetrahedron 58	
3.26	Regular Pyramid 59	
3.27	Frustum of a Regular Pyramid 61	
3.28	Rectangular Right Wedge 62	
3.29	Platonic Solids 63	
3.30	Right Circular Cylinder 66	
3.31	Right Circular Cylinder with an Oblique Plane Face 68	
3.32	Right Circular Cone 69	
3.33	Frustum of a Right Circular Cone 70	
3.34	Sphere 72	
3.35	Spherical Cap 72	
3.36	Spherical Sector 73	
3.37	Spherical Segment 74	
3.38	Spherical Wedge 75	
3.39	Ellipsoid 76	
3.40	Circular Torus 78	

4 TRIGONOMETRY

4.1	Radian and Degree Measures of Angles 80
4.2	Definitions and Graphs of Trigonometric Functions 81
4.3	Signs of Trigonometric Functions 86
4.4	Trigonometric Functions of Common Angles 87
4.5	Most Important Formulas 88

- 4.6 Reduction Formulas **89**
- 4.7 Periodicity of Trigonometric Functions **90**
- 4.8 Relations between Trigonometric Functions **90**
- 4.9 Addition and Subtraction Formulas **91**
- 4.10 Double Angle Formulas **92**
- 4.11 Multiple Angle Formulas **93**
- 4.12 Half Angle Formulas **94**
- 4.13 Half Angle Tangent Identities **94**
- 4.14 Transforming of Trigonometric Expressions to Product **95**
- 4.15 Transforming of Trigonometric Expressions to Sum **97**
- 4.16 Powers of Trigonometric Functions **98**
- 4.17 Graphs of Inverse Trigonometric Functions **99**
- 4.18 Principal Values of Inverse Trigonometric Functions **102**
- 4.19 Relations between Inverse Trigonometric Functions **103**
- 4.20 Trigonometric Equations **106**
- 4.21 Relations to Hyperbolic Functions **106**

5 MATRICES AND DETERMINANTS
- 5.1 Determinants **107**
- 5.2 Properties of Determinants **109**
- 5.3 Matrices **110**
- 5.4 Operations with Matrices **111**
- 5.5 Systems of Linear Equations **114**

6 VECTORS
- 6.1 Vector Coordinates **118**
- 6.2 Vector Addition **120**
- 6.3 Vector Subtraction **122**
- 6.4 Scaling Vectors **122**
- 6.5 Scalar Product **123**
- 6.6 Vector Product **125**
- 6.7 Triple Product **127**

7 ANALYTIC GEOMETRY
- 7.1 One-Dimensional Coordinate System **130**

	Two-Dimensional Coordinate System **131**
7.3	Straight Line in Plane **139**
7.4	Circle **149**
7.5	Ellipse **152**
7.6	Hyperbola **154**
7.7	Parabola **158**
7.8	Three-Dimensional Coordinate System **161**
7.9	Plane **165**
7.10	Straight Line in Space **175**
7.11	Quadric Surfaces **180**
7.12	Sphere **189**

8 DIFFERENTIAL CALCULUS
- 8.1 Functions and Their Graphs **191**
- 8.2 Limits of Functions **208**
- 8.3 Definition and Properties of the Derivative **209**
- 8.4 Table of Derivatives **211**
- 8.5 Higher Order Derivatives **215**
- 8.6 Applications of Derivative **217**
- 8.7 Differential **221**
- 8.8 Multivariable Functions **222**
- 8.9 Differential Operators **225**

9 INTEGRAL CALCULUS
- 9.1 Indefinite Integral **227**
- 9.2 Integrals of Rational Functions **228**
- 9.3 Integrals of Irrational Functions **231**
- 9.4 Integrals of Trigonometric Functions **237**
- 9.5 Integrals of Hyperbolic Functions **241**
- 9.6 Integrals of Exponential and Logarithmic Functions **242**
- 9.7 Reduction Formulas **243**
- 9.8 Definite Integral **247**
- 9.9 Improper Integral **253**
- 9.10 Double Integral **257**
- 9.11 Triple Integral **269**

           Line Integral **275**
  9.13 Surface Integral **285**

10 DIFFERENTIAL EQUATIONS
  10.1 First Order Ordinary Differential Equations **295**
  10.2 Second Order Ordinary Differential Equations **298**
  10.3 Some Partial Differential Equations **302**

11 SERIES
  11.1 Arithmetic Series **304**
  11.2 Geometric Series **305**
  11.3 Some Finite Series **305**
  11.4 Infinite Series **307**
  11.5 Properties of Convergent Series **307**
  11.6 Convergence Tests **308**
  11.7 Alternating Series **310**
  11.8 Power Series **311**
  11.9 Differentiation and Integration of Power Series **312**
  11.10 Taylor and Maclaurin Series **313**
  11.11 Power Series Expansions for Some Functions **314**
  11.12 Binomial Series **316**
  11.13 Fourier Series **316**

12 PROBABILITY
  12.1 Permutations and Combinations **318**
  12.2 Probability Formulas **319**

CHAPTER 1. NUMBER SETS

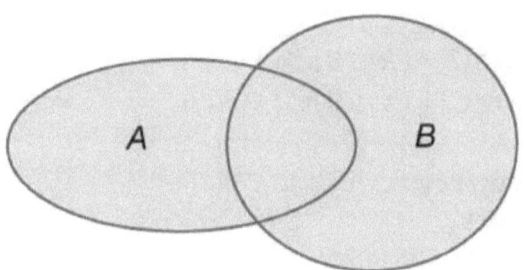

1.1 Set Identities

Figure 1.

6. Commutativity
   $A \cup B = B \cup A$
   Universal set: I
7. Associativity
   $A \cup (B \cup C) = (A \cup B) \cup C$
   Empty set: ∅
8. Intersection of Sets
   $C = A \cup B = \{x \mid x \in A \text{ and } x \in B\}$
   Difference of sets: A \ B

1. $A \subset I$

2. $A \subset A$

3. $A = B$ i:

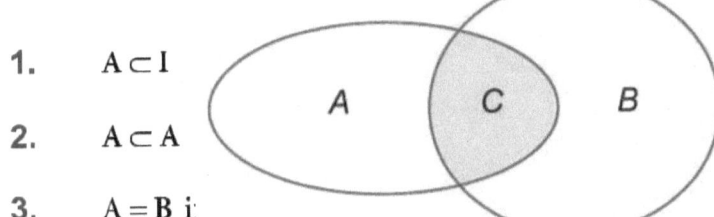

4. Empty Set
   $\emptyset \subset A$

Figure 2.

9.
5. Commutativity
   $A \cap B = B \cap A$
   $C = A \cup B = \{x \mid x \in A \text{ or } x \in B\}$

10. Associativity
    $A \cap (B \cap C) = (A \cap B) \cap C$

11. Distributivity
$$A\cup(B\cap C)=(A\cup B)\cap(A\cup C),$$
$$A\cap(B\cup C)=(A\cap B)\cup(A\cap C).$$

12. Idempotency
$$A\cap A = A,$$
$$A\cup A = A$$

13. Domination
$$A\cap\varnothing=\varnothing,$$
$$A\cup I = I$$

14. Identity
$$A\cup\varnothing = A,$$
$$A\cap I = A$$

15. Complement
$$A' = \{x\in I \mid x\notin A\}$$

16. Complement of Intersection and Union
$$A\cup A' = I,$$
$$A\cap A' = \varnothing$$

17. De Morgan's Laws
$$(A\cup B)' = A'\cap B',$$
$$(A\cap B)' = A'\cup B'$$

18. Difference of Sets
$$C = B\setminus A = \{x \mid x\in B \text{ and } x\notin A\}$$

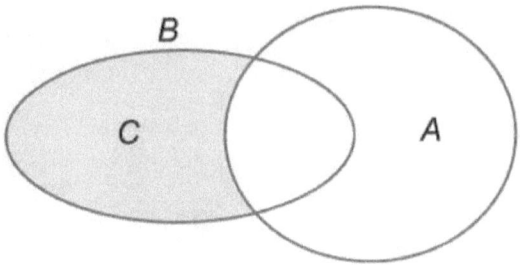

Figure 3.

19. $B \setminus A = B \setminus (A \cap B)$

20. $B \setminus A = B \cap A'$

21. $A \setminus A = \emptyset$

22. $A \setminus B = A$ if $A \cap B = \emptyset$.

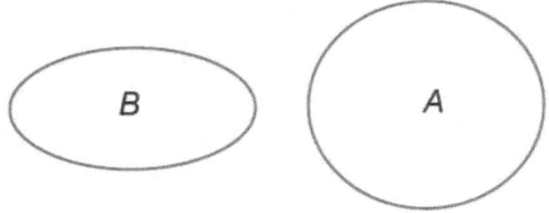

Figure 4.

23. $(A \setminus B) \cap C = (A \cap C) \setminus (B \cap C)$

24. $A' = I \setminus A$

25. Cartesian Product
    $C = A \times B = \{(x, y) | x \in A \text{ and } y \in B\}$

1.2 Sets of Numbers

Natural numbers: N
Whole numbers: N_0
Integers: Z
Positive integers: Z^+
Negative integers: Z^-
Rational numbers: Q
Real numbers: R
Complex numbers: C

26. **Natural Numbers**
    Counting numbers: $N = \{1, 2, 3, ...\}$.

27. **Whole Numbers**
    Counting numbers and zero: $N_0 = \{0, 1, 2, 3, ...\}$.

28. **Integers**
    Whole numbers and their opposites and zero:
    $Z' = N = \{1, 2, 3, ...\}$,
    $Z^- = \{..., -3, -2, -1\}$,
    $Z = Z^- \cup \{0\} \cup Z' = \{..., -3, -2, -1, 0, 1, 2, 3, ...\}$.

29. **Rational Numbers**
    Repeating or terminating decimals:
    $Q = \left\{ x \mid x = \dfrac{a}{b} \text{ and } a \in Z \text{ and } b \in Z \text{ and } b \neq 0 \right\}$.

30. **Irrational Numbers**
    Nonrepeating and nonterminating decimals.

31. Real Numbers
    Union of rational and irrational numbers: R.

32. Complex Numbers
    $C = \{x + iy \mid x \in R \text{ and } y \in R\}$,
    where i is the imaginary unit.

33. $N \subset Z \subset Q \subset R \subset C$

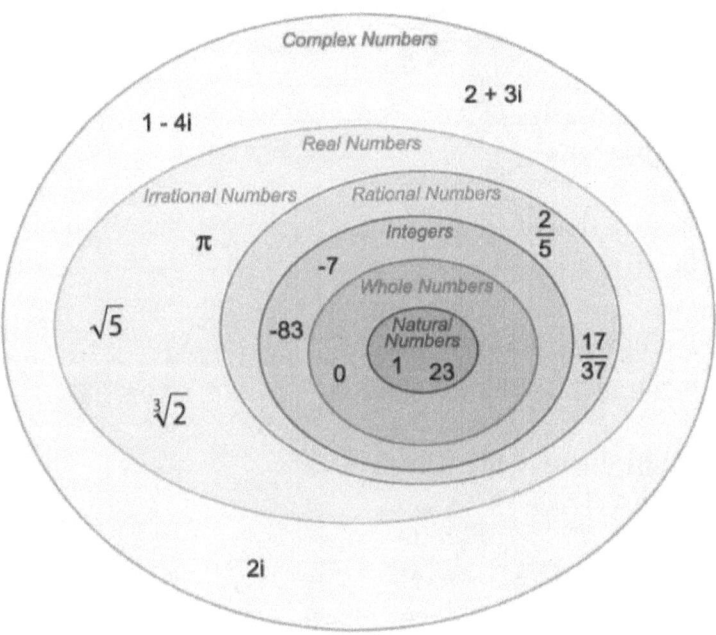

Figure 5.

1.3 Basic Identities

Real numbers: a, b, c

34. Additive Identity
$a + 0 = a$

35. Additive Inverse
$a + (-a) = 0$

36. Commutative of Addition
$a + b = b + a$

37. Associative of Addition
$(a + b) + c = a + (b + c)$

38. Definition of Subtraction
$a - b = a + (-b)$

39. Multiplicative Identity
$a \cdot 1 = a$

40. Multiplicative Inverse
$a \cdot \dfrac{1}{a} = 1, \ a \neq 0$

41. Multiplication Times 0
$a \cdot 0 = 0$

42. Commutative of Multiplication
$a \cdot b = b \cdot a$

43. Associative of Multiplication
$$(a \cdot b) \cdot c = a \cdot (b \cdot c)$$

44. Distributive Law
$$a(b+c) = ab + ac$$

45. Definition of Division
$$\frac{a}{b} = a \cdot \frac{1}{b}$$

1.4 Complex Numbers

Natural number: n
Imaginary unit: i
Complex number: z
Real part: a, c
Imaginary part: bi, di
Modulus of a complex number: r, r_1, r_2
Argument of a complex number: φ, φ_1, φ_2

46.

$i^1 = i$	$i^5 = i$	$i^{4n+1} = i$
$i^2 = -1$	$i^6 = -1$	$i^{4n+2} = -1$
$i^3 = -i$	$i^7 = -i$	$i^{4n+3} = -i$
$i^4 = 1$	$i^8 = 1$	$i^{4n} = 1$

47. $z = a + bi$

48. Complex Plane

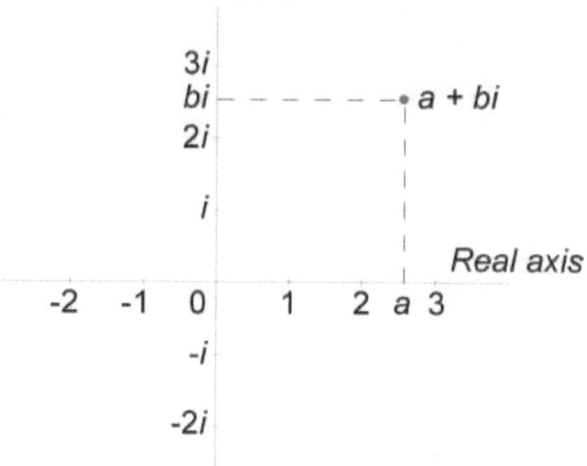

Figure 6.

49. $(a+bi)+(c+di)=(a+c)+(b+d)i$

50. $(a+bi)-(c+di)=(a-c)+(b-d)i$

51. $(a+bi)(c+di)=(ac-bd)+(ad+bc)i$

52. $\dfrac{a+bi}{c+di}=\dfrac{ac+bd}{c^2+d^2}+\dfrac{bc-ad}{c^2+d^2}\cdot i$

53. Conjugate Complex Numbers
    $\overline{a+bi}=a-bi$

54. $a=r\cos\varphi,\ b=r\sin\varphi$

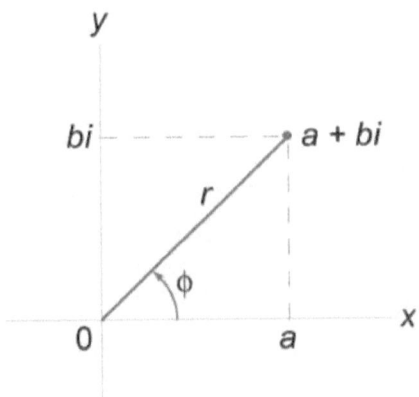

Figure 7.

55. Polar Presentation of Complex Numbers
$$a + bi = r(\cos\varphi + i\sin\varphi)$$

56. Modulus and Argument of a Complex Number
If $a + bi$ is a complex number, then
$$r = \sqrt{a^2 + b^2} \quad \text{(modulus)},$$
$$\varphi = \arctan\frac{b}{a} \quad \text{(argument)}.$$

57. Product in Polar Representation
$$z_1 \cdot z_2 = r_1(\cos\varphi_1 + i\sin\varphi_1) \cdot r_2(\cos\varphi_2 + i\sin\varphi_2)$$
$$= r_1 r_2 [\cos(\varphi_1 + \varphi_2) + i\sin(\varphi_1 + \varphi_2)]$$

58. Conjugate Numbers in Polar Representation
$$\overline{r(\cos\varphi + i\sin\varphi)} = r[\cos(-\varphi) + i\sin(-\varphi)]$$

59. Inverse of a Complex Number in Polar Representation
$$\frac{1}{r(\cos\varphi + i\sin\varphi)} = \frac{1}{r}[\cos(-\varphi) + i\sin(-\varphi)]$$

60. Quotient in Polar Representation
$$\frac{z_1}{z_2} = \frac{r_1(\cos\varphi_1 + i\sin\varphi_1)}{r_2(\cos\varphi_2 + i\sin\varphi_2)} = \frac{r_1}{r_2}[\cos(\varphi_1 - \varphi_2) + i\sin(\varphi_1 - \varphi_2)]$$

61. Power of a Complex Number
$$z^n = [r(\cos\varphi + i\sin\varphi)]^n = r^n[\cos(n\varphi) + i\sin(n\varphi)]$$

62. Formula "De Moivre"
$$(\cos\varphi + i\sin\varphi)^n = \cos(n\varphi) + i\sin(n\varphi)$$

63. Nth Root of a Complex Number
$$\sqrt[n]{z} = \sqrt[n]{r(\cos\varphi + i\sin\varphi)} = \sqrt[n]{r}\left(\cos\frac{\varphi + 2\pi k}{n} + i\sin\frac{\varphi + 2\pi k}{n}\right),$$
where
$k = 0, 1, 2, \ldots, n-1$.

64. Euler's Formula
$$e^{ix} = \cos x + i\sin x$$

Chapter 2
Algebra

2.1 Factoring Formulas

Real numbers: a, b, c
Natural number: n

65. $a^2 - b^2 = (a+b)(a-b)$

66. $a^3 - b^3 = (a-b)(a^2 + ab + b^2)$

67. $a^3 + b^3 = (a+b)(a^2 - ab + b^2)$

68. $a^4 - b^4 = (a^2 - b^2)(a^2 + b^2) = (a-b)(a+b)(a^2 + b^2)$

69. $a^5 - b^5 = (a-b)(a^4 + a^3b + a^2b^2 + ab^3 + b^4)$

70. $a^5 + b^5 = (a+b)(a^4 - a^3b + a^2b^2 - ab^3 + b^4)$

71. If n is odd, then
$a^n + b^n = (a+b)(a^{n-1} - a^{n-2}b + a^{n-3}b^2 - \ldots - ab^{n-2} + b^{n-1})$.

72. If n is even, then
$a^n - b^n = (a-b)(a^{n-1} + a^{n-2}b + a^{n-3}b^2 + \ldots + ab^{n-2} + b^{n-1})$,

$$a^n + b^n = (a+b)(a^{n-1} - a^{n-2}b + a^{n-3}b^2 - \ldots + ab^{n-2} - b^{n-1}).$$

2.2 Product Formulas

Real numbers: a, b, c
Whole numbers: n, k

73. $(a-b)^2 = a^2 - 2ab + b^2$

74. $(a+b)^2 = a^2 + 2ab + b^2$

75. $(a-b)^3 = a^3 - 3a^2b + 3ab^2 - b^3$

76. $(a+b)^3 = a^3 + 3a^2b + 3ab^2 + b^3$

77. $(a-b)^4 = a^4 - 4a^3b + 6a^2b^2 - 4ab^3 + b^4$

78. $(a+b)^4 = a^4 + 4a^3b + 6a^2b^2 + 4ab^3 + b^4$

79. Binomial Formula
$$(a+b)^n = {}^nC_0 a^n + {}^nC_1 a^{n-1}b + {}^nC_2 a^{n-2}b^2 + \ldots + {}^nC_{n-1}ab^{n-1} + {}^nC_n b^n,$$
where ${}^nC_k = \dfrac{n!}{k!(n-k)!}$ are the binomial coefficients.

80. $(a+b+c)^2 = a^2 + b^2 + c^2 + 2ab + 2ac + 2bc$

81. $(a+b+c+\ldots+u+v)^2 = a^2 + b^2 + c^2 + \ldots + u^2 + v^2 +$
$+ 2(ab + ac + \ldots + au + av + bc + \ldots + bu + bv + \ldots + uv)$

2.3 Powers

Bases (positive real numbers): a, b
Powers (rational numbers): n, m

82. $a^m a^n = a^{m+n}$

83. $\dfrac{a^m}{a^n} = a^{m-n}$

84. $(ab)^m = a^m b^m$

85. $\left(\dfrac{a}{b}\right)^m = \dfrac{a^m}{b^m}$

86. $(a^m)^n = a^{mn}$

87. $a^0 = 1, \; a \neq 0$

88. $a^1 = 1$

89. $a^{-m} = \dfrac{1}{a^m}$

90. $a^{\frac{m}{n}} = \sqrt[n]{a^m}$

2.4 Roots

Bases: a, b
Powers (rational numbers): n, m
$a, b \geq 0$ for even roots ($n = 2k$, $k \in \mathbb{N}$)

91. $\sqrt[n]{ab} = \sqrt[n]{a} \sqrt[n]{b}$

92. $\sqrt[n]{a} \sqrt[m]{b} = \sqrt[nm]{a^m b^n}$

93. $\sqrt[n]{\dfrac{a}{b}} = \dfrac{\sqrt[n]{a}}{\sqrt[n]{b}}$, $b \neq 0$

94. $\dfrac{\sqrt[n]{a}}{\sqrt[m]{b}} = \dfrac{\sqrt[nm]{a^m}}{\sqrt[nm]{b^n}} = \sqrt[nm]{\dfrac{a^m}{b^n}}$, $b \neq 0$.

95. $\left(\sqrt[n]{a^m}\right)^p = \sqrt[n]{a^{mp}}$

96. $\left(\sqrt[n]{a}\right)^n = a$

97. $\sqrt[n]{a^m} = \sqrt[np]{a^{mp}}$

98. $\sqrt[n]{a^m} = a^{\frac{m}{n}}$

99. $\sqrt[m]{\sqrt[n]{a}} = \sqrt[mn]{a}$

100. $\left(\sqrt[n]{a}\right)^m = \sqrt[n]{a^m}$

101. $\dfrac{1}{\sqrt[n]{a}} = \dfrac{\sqrt[n]{a^{n-1}}}{a}$, $a \neq 0$.

102. $\sqrt{a \pm \sqrt{b}} = \sqrt{\dfrac{a + \sqrt{a^2 - b}}{2}} \pm \sqrt{\dfrac{a - \sqrt{a^2 - b}}{2}}$

103. $\dfrac{1}{\sqrt{a} \pm \sqrt{b}} = \dfrac{\sqrt{a} \mp \sqrt{b}}{a - b}$

2.5 Logarithms

Positive real numbers: x, y, a, c, k
Natural number: n

104. **Definition of Logarithm**
$y = \log_a x$ if and only if $x = a^y$, $a > 0$, $a \neq 1$.

105. $\log_a 1 = 0$

106. $\log_a a = 1$

107. $\log_a 0 = \begin{cases} -\infty & \text{if } a > 1 \\ +\infty & \text{if } a < 1 \end{cases}$

108. $\log_a (xy) = \log_a x + \log_a y$

109. $\log_a \dfrac{x}{y} = \log_a x - \log_a y$

110. $\log_a(x^n) = n \log_a x$

111. $\log_a \sqrt[n]{x} = \frac{1}{n} \log_a x$

112. $\log_a x = \frac{\log_c x}{\log_c a} = \log_c x \cdot \log_a c$, $c > 0$, $c \neq 1$.

113. $\log_a c = \frac{1}{\log_c a}$

114. $x = a^{\log_a x}$

115. **Logarithm to Base 10**
$\log_{10} x = \log x$

116. **Natural Logarithm**
$\log_e x = \ln x$,
where $e = \lim\limits_{k \to \infty} \left(1 + \frac{1}{k}\right)^k = 2.718281828\ldots$

117. $\log x = \frac{1}{\ln 10} \ln x = 0.434294 \ln x$

118. $\ln x = \frac{1}{\log e} \log x = 2.302585 \log x$

2.6 Equations

Real numbers: a, b, c, p, q, u, v
Solutions: x_1, x_2, y_1, y_2, y_3

119. Linear Equation in One Variable
$ax + b = 0$, $x = -\dfrac{b}{a}$.

120. Quadratic Equation
$ax^2 + bx + c = 0$, $x_{1,2} = \dfrac{-b \pm \sqrt{b^2 - 4ac}}{2a}$.

121. Discriminant
$D = b^2 - 4ac$

122. Viete's Formulas
If $x^2 + px + q = 0$, then
$\begin{cases} x_1 + x_2 = -p \\ x_1 x_2 = q \end{cases}$.

123. $ax^2 + bx = 0$, $x_1 = 0$, $x_2 = -\dfrac{b}{a}$.

124. $ax^2 + c = 0$, $x_{1,2} = \pm\sqrt{-\dfrac{c}{a}}$.

125. Cubic Equation. Cardano's Formula.
$y^3 + py + q = 0$,

$$y_1 = u+v, \quad y_{2,3} = -\frac{1}{2}(u+v) \pm \frac{\sqrt{3}}{2}(u+v)i,$$

where

$$u = \sqrt[3]{-\frac{q}{2} + \sqrt{\left(\frac{q}{2}\right)^2 + \left(\frac{p}{3}\right)^2}}, \quad v = \sqrt[3]{-\frac{q}{2} - \sqrt{\left(\frac{q}{2}\right)^2 + \left(\frac{p}{3}\right)^2}}.$$

2.7 Inequalities

Variables: x, y, z

Real numbers: $\begin{cases} a, b, c, d \\ a_1, a_2, a_3, \ldots, a_n \end{cases}$, m, n

Determinants: D, D_x, D_y, D_z

126. Inequalities, Interval Notations and Graphs

Inequality	Interval Notation	Graph
$a \leq x \leq b$	$[a, b]$	•——•→ x (a, b)
$a < x \leq b$	$(a, b]$	○——•→ x (a, b)
$a \leq x < b$	$[a, b)$	•——○→ x (a, b)
$a < x < b$	(a, b)	○——○→ x (a, b)
$-\infty < x \leq b$, $x \leq b$	$(-\infty, b]$	——————•→ x (b)
$-\infty < x < b$, $x < b$	$(-\infty, b)$	——————○→ x (b)
$a \leq x < \infty$, $x \geq a$	$[a, \infty)$	•——————→ x (a)
$a < x < \infty$, $x > a$	(a, ∞)	○——————→ x (a)

CHAPTER 2. ALGEBRA

127. If $a > b$, then $b < a$.

128. If $a > b$, then $a - b > 0$ or $b - a < 0$.

129. If $a > b$, then $a + c > b + c$.

130. If $a > b$, then $a - c > b - c$.

131. If $a > b$ and $c > d$, then $a + c > b + d$.

132. If $a > b$ and $c > d$, then $a - d > b - c$.

133. If $a > b$ and $m > 0$, then $ma > mb$.

134. If $a > b$ and $m > 0$, then $\dfrac{a}{m} > \dfrac{b}{m}$.

135. If $a > b$ and $m < 0$, then $ma < mb$.

136. If $a > b$ and $m < 0$, then $\dfrac{a}{m} < \dfrac{b}{m}$.

137. If $0 < a < b$ and $n > 0$, then $a^n < b^n$.

138. If $0 < a < b$ and $n < 0$, then $a^n > b^n$.

139. If $0 < a < b$, then $\sqrt[n]{a} < \sqrt[n]{b}$.

140. $\sqrt{ab} \leq \dfrac{a+b}{2}$,
where $a > 0$, $b > 0$; an equality is valid only if $a = b$.

141. $a + \dfrac{1}{a} \geq 2$, where $a > 0$; an equality takes place only at $a = 1$.

CHAPTER 2. ALGEBRA

142. $\sqrt[n]{a_1 a_2 \ldots a_n} \leq \dfrac{a_1 + a_2 + \ldots + a_n}{n}$, where $a_1, a_2, \ldots, a_n > 0$.

143. If $ax + b > 0$ and $a > 0$, then $x > -\dfrac{b}{a}$.

144. If $ax + b > 0$ and $a < 0$, then $x < -\dfrac{b}{a}$.

145. $ax^2 + bx + c > 0$

	$a > 0$	$a < 0$
$D > 0$	$x < x_1, \ x > x_2$	$x_1 < x < x_2$
$D = 0$	$x_1 < x, \ x > x_1$	$x \in \varnothing$
$D < 0$	$-\infty < x < \infty$	$x \in \varnothing$

146. $|a+b| \le |a|+|b|$

147. If $|x| < a$, then $-a < x < a$, where $a > 0$.

148. If $|x| > a$, then $x < -a$ and $x > a$, where $a > 0$.

149. If $x^2 < a$, then $|x| < \sqrt{a}$, where $a > 0$.

150. If $x^2 > a$, then $|x| > \sqrt{a}$, where $a > 0$.

151. If $\dfrac{f(x)}{g(x)} > 0$, then $\begin{cases} f(x) \cdot g(x) > 0 \\ g(x) \ne 0 \end{cases}$.

152. $\dfrac{f(x)}{g(x)} < 0$, then $\begin{cases} f(x) \cdot g(x) < 0 \\ g(x) \ne 0 \end{cases}$.

2.8 Compound Interest Formulas

Future value: A
Initial deposit: C
Annual rate of interest: r
Number of years invested: t
Number of times compounded per year: n

153. General Compound Interest Formula
$$A = C\left(1 + \frac{r}{n}\right)^{nt}$$

154. **Simplified Compound Interest Formula**
If interest is compounded once per year, then the previous formula simplifies to:
$A = C(1+r)^t$.

155. **Continuous Compound Interest**
If interest is compounded continually ($n \to \infty$), then
$A = Ce^{rt}$.

Chapter 3

Geometry

3.1 Right Triangle

Legs of a right triangle: a, b
Hypotenuse: c
Altitude: h
Medians: m_a, m_b, m_c
Angles: α, β
Radius of circumscribed circle: R
Radius of inscribed circle: r
Area: S

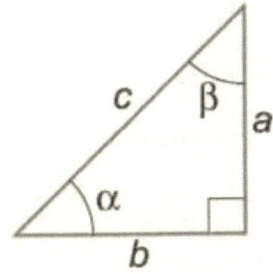

Figure 8.

156. $\alpha + \beta = 90°$

CHAPTER 3. GEOMETRY

157. $\sin \alpha = \dfrac{a}{c} = \cos \beta$

158. $\cos \alpha = \dfrac{b}{c} = \sin \beta$

159. $\tan \alpha = \dfrac{a}{b} = \cot \beta$

160. $\cot \alpha = \dfrac{b}{a} = \tan \beta$

161. $\sec \alpha = \dfrac{c}{b} = \operatorname{cosec} \beta$

162. $\operatorname{cosec} \alpha = \dfrac{c}{a} = \sec \beta$

163. Pythagorean Theorem
$a^2 + b^2 = c^2$

164. $a^2 = fc$, $b^2 = gc$,
where f and c are projections of the legs a and b, respectively, onto the hypotenuse c.

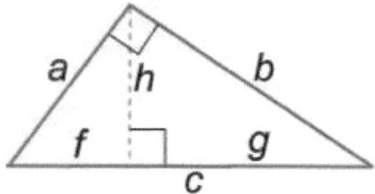

Figure 9.

165. $h^2 = fg$,

where h is the altitude from the right angle.

166. $m_a^2 = b^2 - \dfrac{a^2}{4}$, $m_b^2 = a^2 - \dfrac{b^2}{4}$,

where m_a and m_b are the medians to the legs a and b.

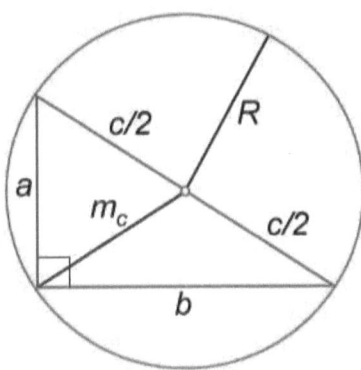

Figure 10.

167. $m_c = \dfrac{c}{2}$,

where m_c is the median to the hypotenuse c.

168. $R = \dfrac{c}{2} = m_c$

169. $r = \dfrac{a+b-c}{2} = \dfrac{ab}{a+b+c}$

170. $ab = ch$

171. $S = \dfrac{ab}{2} = \dfrac{ch}{2}$

3.2 Isosceles Triangle

Base: a
Legs: b
Base angle: β
Vertex angle: α
Altitude to the base: h
Perimeter: L
Area: S

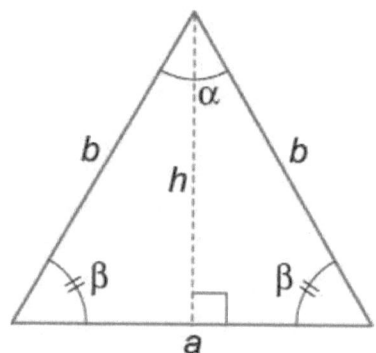

Figure 11.

172. $\beta = 90° - \dfrac{\alpha}{2}$

173. $h^2 = b^2 - \dfrac{a^2}{4}$

174. $L = a + 2b$

175. $S = \dfrac{ah}{2} = \dfrac{b^2}{2}\sin\alpha$

3.3 Equilateral Triangle

Side of a equilateral triangle: a
Altitude: h
Radius of circumscribed circle: R
Radius of inscribed circle: r
Perimeter: L
Area: S

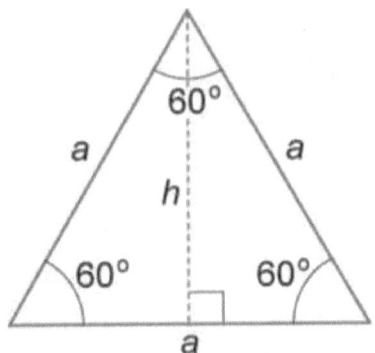

Figure 12.

176. $h = \dfrac{a\sqrt{3}}{2}$

177. $R = \dfrac{2}{3}h = \dfrac{a\sqrt{3}}{3}$

178. $r = \dfrac{1}{3}h = \dfrac{a\sqrt{3}}{6} = \dfrac{R}{2}$

179. $L = 3a$

180. $S = \dfrac{ah}{2} = \dfrac{a^2\sqrt{3}}{4}$

3.4 Scalene Triangle

(A triangle with no two sides equal)

Sides of a triangle: a, b, c
Semiperimeter: $p = \dfrac{a+b+c}{2}$
Angles of a triangle: α, β, γ
Altitudes to the sides a, b, c: h_a, h_b, h_c
Medians to the sides a, b, c: m_a, m_b, m_c
Bisectors of the angles α, β, γ: t_a, t_b, t_c
Radius of circumscribed circle: R
Radius of inscribed circle: r
Area: S

CHAPTER 3. GEOMETRY

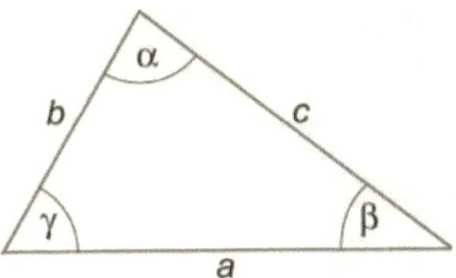

Figure 13.

181. $\alpha + \beta + \gamma = 180°$

182. $a + b > c$,
$b + c > a$,
$a + c > b$.

183. $|a - b| < c$,
$|b - c| < a$,
$|a - c| < b$.

184. Midline
$q = \dfrac{a}{2}$, $q \| a$.

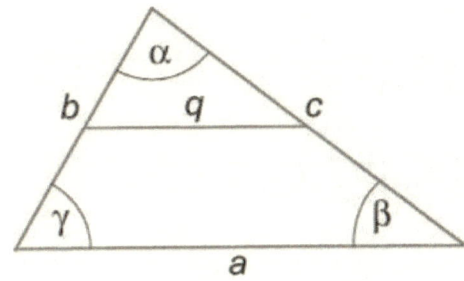

Figure 14.

CHAPTER 3. GEOMETRY

185. Law of Cosines
$$a^2 = b^2 + c^2 - 2bc \cos \alpha,$$
$$b^2 = a^2 + c^2 - 2ac \cos \beta,$$
$$c^2 = a^2 + b^2 - 2ab \cos \gamma.$$

186. Law of Sines
$$\frac{a}{\sin \alpha} = \frac{b}{\sin \beta} = \frac{c}{\sin \gamma} = 2R,$$
where R is the radius of the circumscribed circle.

187. $R = \dfrac{a}{2 \sin \alpha} = \dfrac{b}{2 \sin \beta} = \dfrac{c}{2 \sin \gamma} = \dfrac{bc}{2h_a} = \dfrac{ac}{2h_b} = \dfrac{ab}{2h_c} = \dfrac{abc}{4S}$

188. $r^2 = \dfrac{(p-a)(p-b)(p-c)}{p},$

$\dfrac{1}{r} = \dfrac{1}{h_a} + \dfrac{1}{h_b} + \dfrac{1}{h_c}.$

189. $\sin \dfrac{\alpha}{2} = \sqrt{\dfrac{(p-b)(p-c)}{bc}},$

$\cos \dfrac{\alpha}{2} = \sqrt{\dfrac{p(p-a)}{bc}},$

$\tan \dfrac{\alpha}{2} = \sqrt{\dfrac{(p-b)(p-c)}{p(p-a)}}.$

190. $h_a = \dfrac{2}{a}\sqrt{p(p-a)(p-b)(p-c)},$

$h_b = \dfrac{2}{b}\sqrt{p(p-a)(p-b)(p-c)},$

$h_c = \dfrac{2}{c}\sqrt{p(p-a)(p-b)(p-c)}.$

CHAPTER 3. GEOMETRY

191. $h_a = b\sin\gamma = c\sin\beta$,
$h_b = a\sin\gamma = c\sin\alpha$,
$h_c = a\sin\beta = b\sin\alpha$.

192. $m_a^2 = \dfrac{b^2+c^2}{2} - \dfrac{a^2}{4}$,

$m_b^2 = \dfrac{a^2+c^2}{2} - \dfrac{b^2}{4}$,

$m_c^2 = \dfrac{a^2+b^2}{2} - \dfrac{c^2}{4}$.

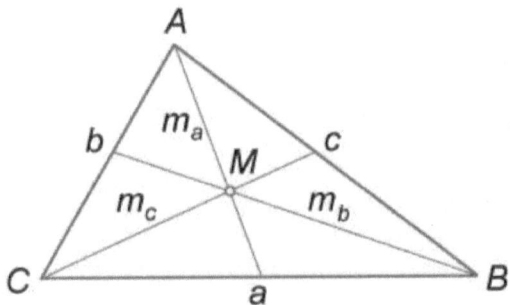

Figure 15.

193. $AM = \dfrac{2}{3}m_a$, $BM = \dfrac{2}{3}m_b$, $CM = \dfrac{2}{3}m_c$ (Fig.15).

194. $t_a^2 = \dfrac{4bcp(p-a)}{(b+c)^2}$,

$t_b^2 = \dfrac{4acp(p-b)}{(a+c)^2}$,

$t_c^2 = \dfrac{4abp(p-c)}{(a+b)^2}$.

195. $S = \dfrac{ah_a}{2} = \dfrac{bh_b}{2} = \dfrac{ch_c}{2}$,

$S = \dfrac{ab \sin \gamma}{2} = \dfrac{ac \sin \beta}{2} = \dfrac{bc \sin \alpha}{2}$,

$S = \sqrt{p(p-a)(p-b)(p-c)}$ (Heron's Formula),

$S = pr$,

$S = \dfrac{abc}{4R}$,

$S = 2R^2 \sin \alpha \sin \beta \sin \gamma$,

$S = p^2 \tan \dfrac{\alpha}{2} \tan \dfrac{\beta}{2} \tan \dfrac{\gamma}{2}$.

3.5 Square

Side of a square: a
Diagonal: d
Radius of circumscribed circle: R
Radius of inscribed circle: r
Perimeter: L
Area: S

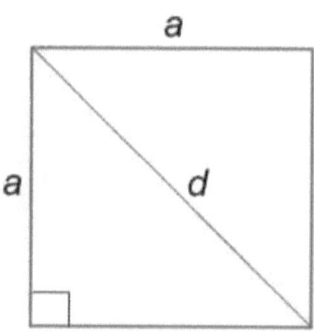

Figure 16.

196. $d = a\sqrt{2}$

197. $R = \dfrac{d}{2} = \dfrac{a\sqrt{2}}{2}$

198. $r = \dfrac{a}{2}$

199. $L = 4a$

200. $S = a^2$

3.6 Rectangle

Sides of a rectangle: a, b
Diagonal: d
Radius of circumscribed circle: R
Perimeter: L
Area: S

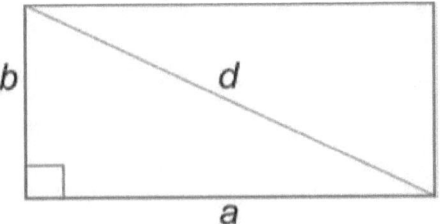

Figure 17.

201. $d = \sqrt{a^2 + b^2}$

202. $R = \dfrac{d}{2}$

203. $L = 2(a + b)$

204. $S = ab$

3.7 Parallelogram

Sides of a parallelogram: a, b
Diagonals: d_1, d_2
Consecutive angles: α, β
Angle between the diagonals: φ
Altitude: h
Perimeter: L
Area: S

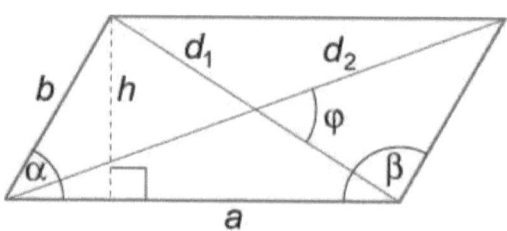

Figure 18.

205. $\alpha + \beta = 180°$

206. $d_1^2 + d_2^2 = 2(a^2 + b^2)$

207. $h = b \sin \alpha = b \sin \beta$

208. $L = 2(a + b)$

209. $S = ah = ab \sin \alpha$,

$S = \dfrac{1}{2} d_1 d_2 \sin \varphi$.

3.8 Rhombus

Side of a rhombus: a
Diagonals: d_1, d_2
Consecutive angles: α, β
Altitude: H
Radius of inscribed circle: r
Perimeter: L
Area: S

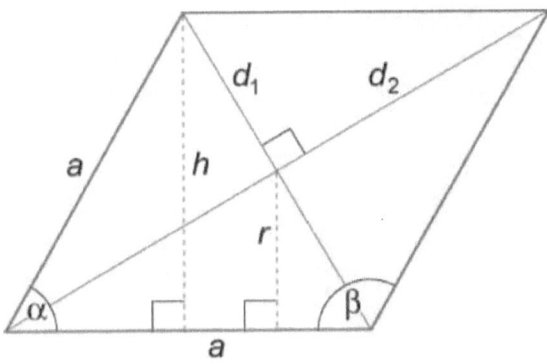

Figure 19.

210. $\alpha + \beta = 180°$

211. $d_1^2 + d_2^2 = 4a^2$

212. $h = a \sin \alpha = \dfrac{d_1 d_2}{2a}$

213. $r = \dfrac{h}{2} = \dfrac{d_1 d_2}{4a} = \dfrac{a \sin \alpha}{2}$

214. $L = 4a$

215. $S = ah = a^2 \sin \alpha$,

$S = \dfrac{1}{2} d_1 d_2$.

3.9 Trapezoid

Bases of a trapezoid: a, b
Midline: q
Altitude: h
Area: S

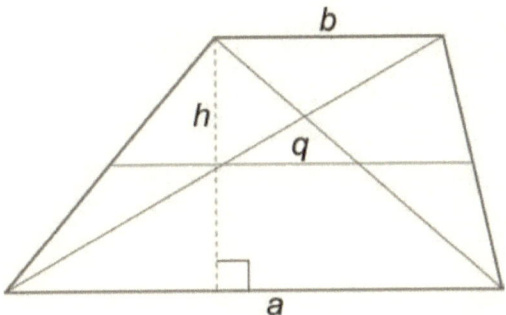

Figure 20.

216. $q = \dfrac{a+b}{2}$

217. $S = \dfrac{a+b}{2} \cdot h = qh$

3.10 Isosceles Trapezoid

Bases of a trapezoid: a, b
Leg: c
Midline: q
Altitude: h
Diagonal: d
Radius of circumscribed circle: R
Area: S

CHAPTER 3. GEOMETRY

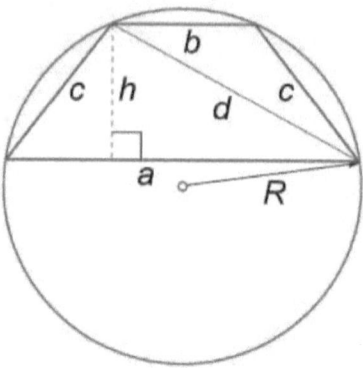

Figure 21.

218. $q = \dfrac{a+b}{2}$

219. $d = \sqrt{ab + c^2}$

220. $h = \sqrt{c^2 - \dfrac{1}{4}(b-a)^2}$

221. $R = \dfrac{c\sqrt{ab+c^2}}{\sqrt{(2c-a+b)(2c+a-b)}}$

222. $S = \dfrac{a+b}{2} \cdot h = qh$

3.11 Isosceles Trapezoid with Inscribed Circle

Bases of a trapezoid: a, b
Leg: c
Midline: q
Altitude: h
Diagonal: d
Radius of inscribed circle: R
Radius of circumscribed circle: r
Perimeter: L
Area: S

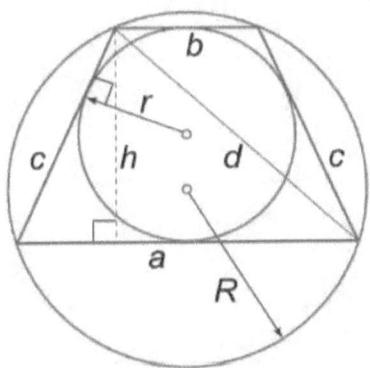

Figure 22.

223. $a + b = 2c$

224. $q = \dfrac{a+b}{2} = c$

225. $d^2 = h^2 + c^2$

226. $r = \dfrac{h}{2} = \dfrac{\sqrt{ab}}{2}$

227. $R = \dfrac{cd}{2h} = \dfrac{cd}{4r} = \dfrac{c}{2}\sqrt{1 + \dfrac{c^2}{ab}} = \dfrac{c}{2h}\sqrt{h^2 + c^2} = \dfrac{a+b}{8}\sqrt{\dfrac{a}{b} + 6 + \dfrac{b}{a}}$

228. $L = 2(a + b) = 4c$

229. $S = \dfrac{a+b}{2} \cdot h = \dfrac{(a+b)\sqrt{ab}}{2} = qh = ch = \dfrac{Lr}{2}$

3.12 Trapezoid with Inscribed Circle

Bases of a trapezoid: a, b
Lateral sides: c, d
Midline: q
Altitude: h
Diagonals: d_1, d_2
Angle between the diagonals: φ
Radius of inscribed circle: r
Radius of circumscribed circle: R
Perimeter: L
Area: S

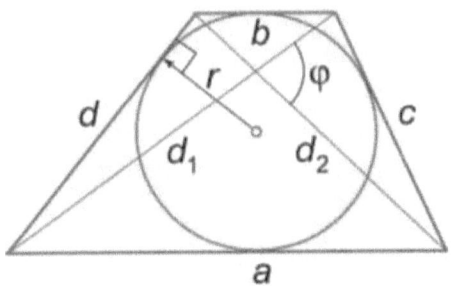

Figure 23.

230. $a + b = c + d$

231. $q = \dfrac{a+b}{2} = \dfrac{c+d}{2}$

232. $L = 2(a+b) = 2(c+d)$

233. $S = \dfrac{a+b}{2} \cdot h = \dfrac{c+d}{2} \cdot h = qh$,

$S = \dfrac{1}{2} d_1 d_2 \sin \varphi$.

3.13 Kite

Sides of a kite: a, b
Diagonals: d_1, d_2
Angles: α, β, γ
Perimeter: L
Area: S

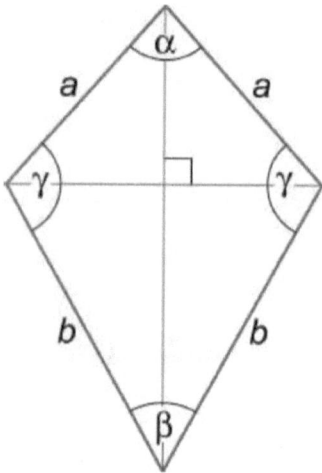

Figure 24.

234. $\alpha + \beta + 2\gamma = 360°$

235. $L = 2(a + b)$

236. $S = \dfrac{d_1 d_2}{2}$

3.14 Cyclic Quadrilateral

Sides of a quadrilateral: a, b, c, d
Diagonals: d_1, d_2
Angle between the diagonals: φ
Internal angles: $\alpha, \beta, \gamma, \delta$
Radius of circumscribed circle: R
Perimeter: L
Semiperimeter: p
Area: S

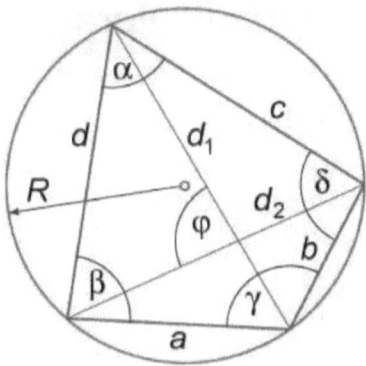

Figure 25.

237. $\alpha + \gamma = \beta + \delta = 180°$

238. Ptolemy's Theorem
$ac + bd = d_1 d_2$

239. $L = a + b + c + d$

240. $R = \dfrac{1}{4}\sqrt{\dfrac{(ac+bd)(ad+bc)(ab+cd)}{(p-a)(p-b)(p-c)(p-d)}}$,

where $p = \dfrac{L}{2}$.

241. $S = \dfrac{1}{2} d_1 d_2 \sin\varphi$,

$S = \sqrt{(p-a)(p-b)(p-c)(p-d)}$,

where $p = \dfrac{L}{2}$.

3.15 Tangential Quadrilateral

Sides of a quadrilateral: a, b, c, d
Diagonals: d_1, d_2
Angle between the diagonals: φ
Radius of inscribed circle: r
Perimeter: L
Semiperimeter: p
Area: S

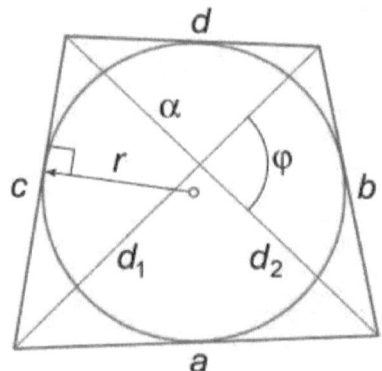

Figure 26.

242. $a + c = b + d$

243. $L = a + b + c + d = 2(a + c) = 2(b + d)$

244. $r = \dfrac{\sqrt{d_1^2 d_2^2 - (a-b)^2(a+b-p)^2}}{2p}$,

where $p = \dfrac{L}{2}$.

245. $S = pr = \dfrac{1}{2} d_1 d_2 \sin \varphi$

3.16 General Quadrilateral

Sides of a quadrilateral: a, b, c, d
Diagonals: d_1, d_2
Angle between the diagonals: φ
Internal angles: $\alpha, \beta, \gamma, \delta$
Perimeter: L
Area: S

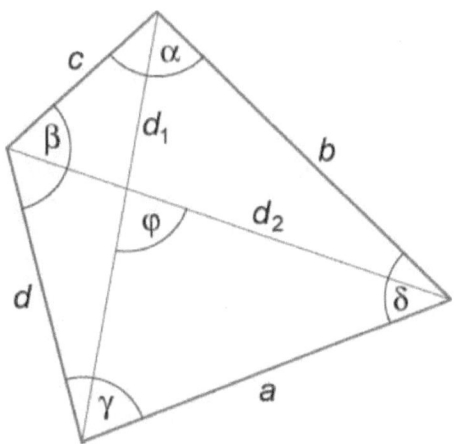

Figure 27.

246. $\alpha + \beta + \gamma + \delta = 360°$

247. $L = a + b + c + d$

248. $S = \dfrac{1}{2} d_1 d_2 \sin \varphi$

3.17 Regular Hexagon

Side: a
Internal angle: α
Slant height: m
Radius of inscribed circle: r
Radius of circumscribed circle: R
Perimeter: L
Semiperimeter: p
Area: S

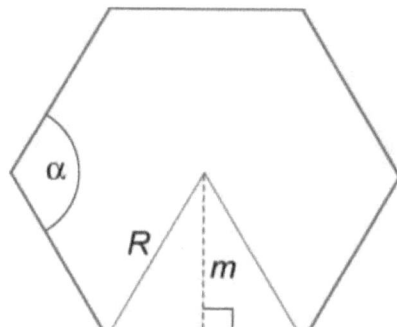

Figure 28.

249. $\alpha = 120°$

250. $r = m = \dfrac{a\sqrt{3}}{2}$

251. $R = a$

252. $L = 6a$

253. $S = pr = \dfrac{a^2 3\sqrt{3}}{2}$,

where $p = \dfrac{L}{2}$.

3.18 Regular Polygon

Side: a
Number of sides: n
Internal angle: α
Slant height: m
Radius of inscribed circle: r
Radius of circumscribed circle: R
Perimeter: L
Semiperimeter: p
Area: S

CHAPTER 3. GEOMETRY

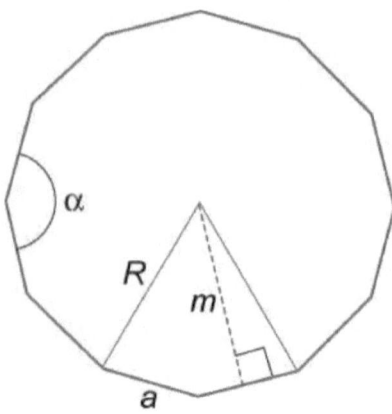

Figure 29.

254. $\alpha = \dfrac{n-2}{2} \cdot 180°$

255. $\alpha = \dfrac{n-2}{2} \cdot 180°$

256. $R = \dfrac{a}{2\sin\dfrac{\pi}{n}}$

257. $r = m = \dfrac{a}{2\tan\dfrac{\pi}{n}} = \sqrt{R^2 - \dfrac{a^2}{4}}$

258. $L = na$

259. $S = \dfrac{nR^2}{2}\sin\dfrac{2\pi}{n},$

$S = pr = p\sqrt{R^2 - \dfrac{a^2}{4}},$

where $p = \dfrac{L}{2}$.

3.19 Circle

Radius: R
Diameter: d
Chord: a
Secant segments: e, f
Tangent segment: g
Central angle: α
Inscribed angle: β
Perimeter: L
Area: S

260. $a = 2R \sin \dfrac{\alpha}{2}$

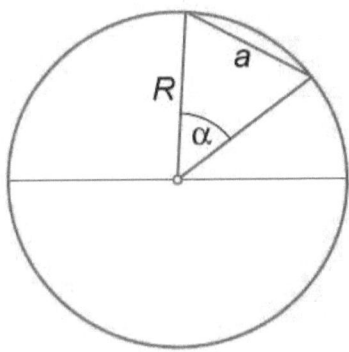

Figure 30.

261. $a_1 a_2 = b_1 b_2$

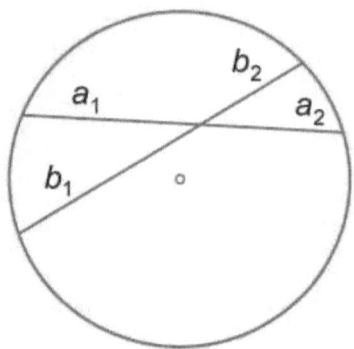

Figure 31.

262. $ee_1 = ff_1$

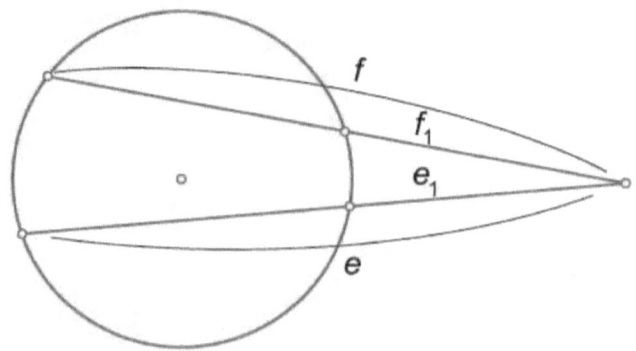

Figure 32.

263. $g^2 = ff_1$

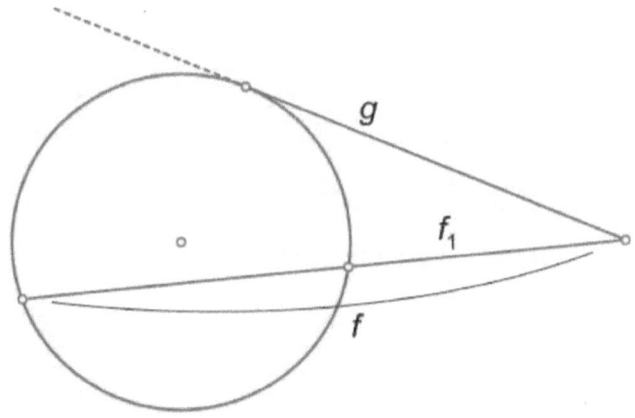

Figure 33.

264. $\beta = \dfrac{\alpha}{2}$

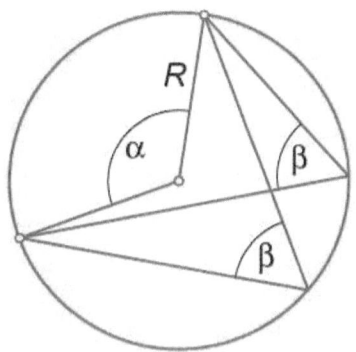

Figure 34.

265. $L = 2\pi R = \pi d$

266. $S = \pi R^2 = \dfrac{\pi d^2}{4} = \dfrac{}{2}$

3.20 Sector of a Circle

Radius of a circle: R
Arc length: s
Central angle (in radians): x
Central angle (in degrees): α
Perimeter: L
Area: S

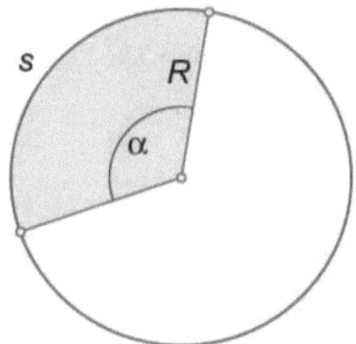

Figure 35.

267. $s = Rx$

268. $s = \dfrac{\pi R \alpha}{180°}$

269. $L = s + 2R$

270. $S = \dfrac{Rs}{2} = \dfrac{R^2 x}{2} = \dfrac{\pi R^2 \alpha}{360°}$

3.21 Segment of a Circle

Radius of a circle: R
Arc length: s
Chord: a
Central angle (in radians): x
Central angle (in degrees): α
Height of the segment: h
Perimeter: L
Area: S

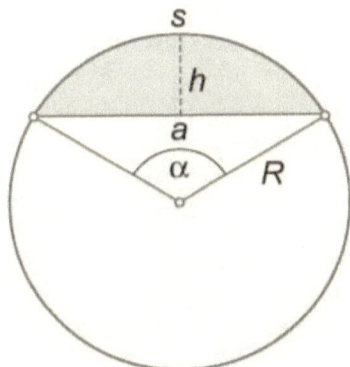

Figure 36.

271. $a = 2\sqrt{2hR - h^2}$

272. $h = R - \frac{1}{2}\sqrt{4R^2 - a^2}$, $h < R$

273. $L = s + a$

274. $S = \frac{1}{2}[sR - a(R-h)] = \frac{R^2}{2}\left(\frac{\alpha\pi}{180°} - \sin\alpha\right) = \frac{R^2}{2}(x - \sin x),$

$S \approx \frac{2}{3}ha.$

3.22 Cube

Edge: a
Diagonal: d
Radius of inscribed sphere: r
Radius of circumscribed sphere: r
Surface area: S
Volume: V

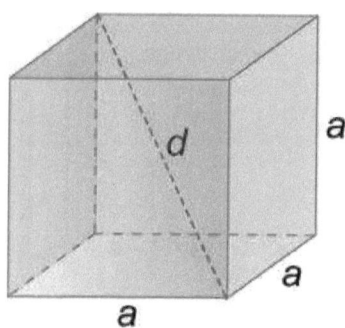

Figure 37.

275. $d = a\sqrt{3}$

276. $r = \frac{a}{2}$

277. $R = \dfrac{a\sqrt{3}}{2}$

278. $S = 6a^2$

279. $V = a^3$

3.23 Rectangular Parallelepiped

Edges: a, b, c
Diagonal: d
Surface area: S
Volume: V

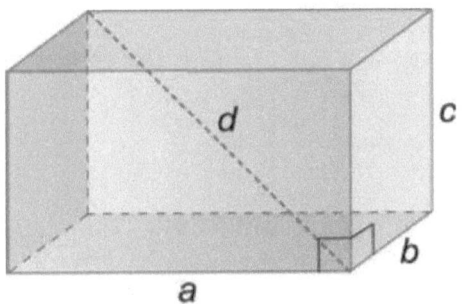

Figure 38.

280. $d = \sqrt{a^2 + b^2 + c^2}$

281. $S = 2(ab + ac + bc)$

282. $V = abc$

3.24 Prism

Lateral edge: l
Height: h
Lateral area: S_L
Area of base: S_B
Total surface area: S
Volume: V

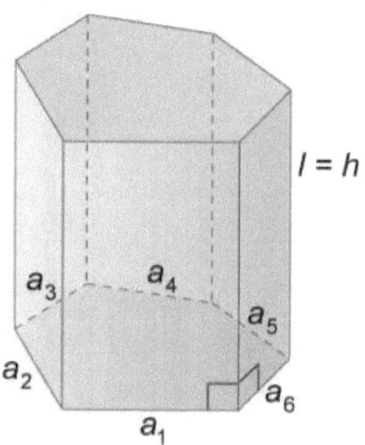

Figure 39.

283. $S = S_L + 2S_B$.

284. Lateral Area of a Right Prism
$S_L = (a_1 + a_2 + a_3 + \ldots + a_n)l$

285. Lateral Area of an Oblique Prism
$S_L = pl$,
where p is the perimeter of the cross section.

CHAPTER 3. GEOMETRY

286. $V = S_B h$

287. Cavalieri's Principle
Given two solids included between parallel planes. If every plane cross section parallel to the given planes has the same area in both solids, then the volumes of the solids are equal.

3.25 Regular Tetrahedron

Triangle side length: a
Height: h
Area of base: S_B
Surface area: S
Volume: V

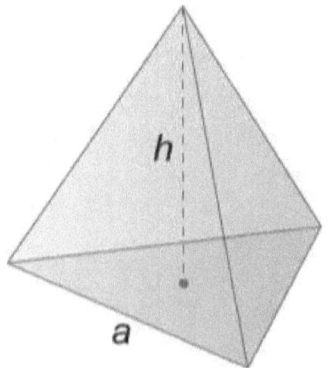

Figure 40.

288. $h = \sqrt{\dfrac{2}{3}}\, a$

289. $S_B = \dfrac{\sqrt{3}a^2}{4}$

290. $S = \sqrt{3}a^2$

291. $V = \dfrac{1}{3}S_B h = \dfrac{a^3}{6\sqrt{2}}$.

3.26 Regular Pyramid

Side of base: a
Lateral edge: b
Height: h
Slant height: m
Number of sides: n
Semiperimeter of base: p
Radius of inscribed sphere of base: r
Area of base: S_B
Lateral surface area: S_L
Total surface area: S
Volume: V

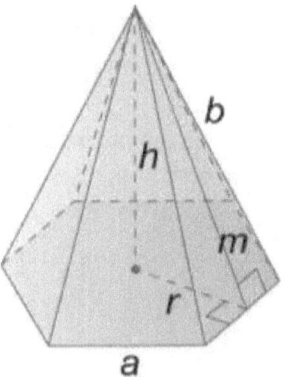

Figure 41.

292. $m = \sqrt{b^2 - \dfrac{a^2}{4}}$

293. $h = \dfrac{\sqrt{4b^2 \sin^2 \dfrac{\pi}{n} - a^2}}{2\sin \dfrac{\pi}{n}}$

294. $S_L = \dfrac{1}{2}nam = \dfrac{1}{4}na\sqrt{4b^2 - a^2} = pm$

295. $S_B = pr$

296. $S = S_B + S_L$

297. $V = \dfrac{1}{3}S_B h = \dfrac{1}{3}prh$

3.27 Frustum of a Regular Pyramid

Base and top side lengths: $\begin{cases} a_1, a_2, a_3, \ldots, a_n \\ b_1, b_2, b_3, \ldots, b_n \end{cases}$

Height: h
Slant height: m
Area of bases: S_1, S_2
Lateral surface area: S_L
Perimeter of bases: P_1, P_2
Scale factor: k
Total surface area: S
Volume: V

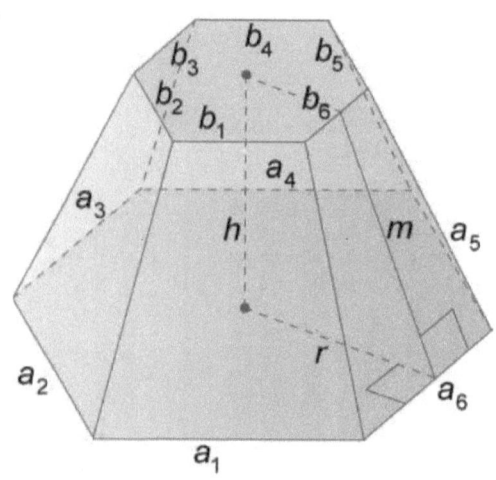

Figure 42.

298. $\dfrac{b_1}{a_1} = \dfrac{b_2}{a_2} = \dfrac{b_3}{a_3} = \ldots = \dfrac{b_n}{a_n} = \dfrac{b}{a} = k$

299. $\dfrac{S_2}{S_1} = k^2$

300. $S_L = \dfrac{m(P_1 + P_2)}{2}$

301. $S = S_L + S_1 + S_2$

302. $V = \dfrac{h}{3}\left(S_1 + \sqrt{S_1 S_2} + S_2\right)$

303. $V = \dfrac{hS_1}{3}\left[1 + \dfrac{b}{a} + \left(\dfrac{b}{a}\right)^2\right] = \dfrac{hS_1}{3}\left[1 + k + k^2\right]$

3.28 Rectangular Right Wedge

Sides of base: a, b
Top edge: c
Height: h
Lateral surface area: S_L
Area of base: S_B
Total surface area: S
Volume: V

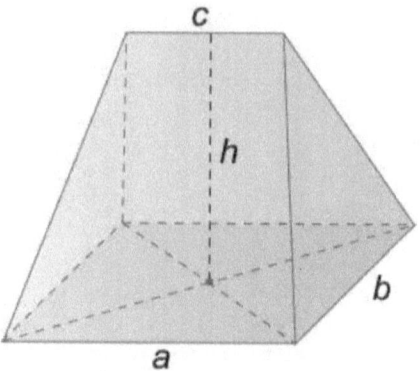

Figure 43.

304. $S_L = \frac{1}{2}(a+c)\sqrt{4h^2+b^2} + b\sqrt{h^2+(a-c)^2}$

305. $S_B = ab$

306. $S = S_B + S_L$

307. $V = \frac{bh}{6}(2a+c)$

3.29 Platonic Solids

Edge: a
Radius of inscribed circle: r
Radius of circumscribed circle: R
Surface area: S
Volume: V

308. Five Platonic Solids

The platonic solids are convex polyhedra with equivalent faces composed of congruent convex regular polygons.

Solid	Number of Vertices	Number of Edges	Number of Faces	Section
Tetrahedron	4	6	4	3.25
Cube	8	12	6	3.22
Octahedron	6	12	8	3.27
Icosahedron	12	30	20	3.27
Dodecahedron	20	30	12	3.27

Octahedron

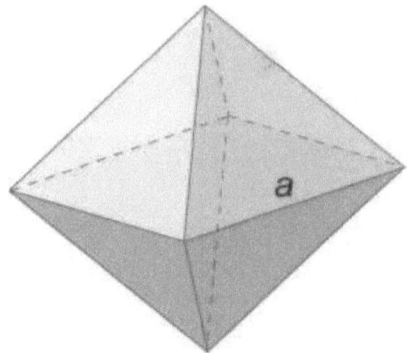

Figure 44.

309. $r = \dfrac{a\sqrt{6}}{6}$

310. $R = \dfrac{a\sqrt{2}}{2}$

311. $S = 2a^2\sqrt{3}$

312. $V = \dfrac{a^3\sqrt{2}}{3}$

Icosahedron

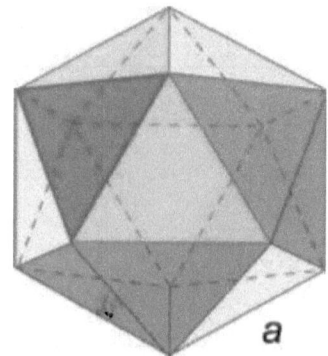

Figure 45.

313. $r = \dfrac{a\sqrt{3}\left(3+\sqrt{5}\right)}{12}$

314. $R = \dfrac{a}{4}\sqrt{2\left(5+\sqrt{5}\right)}$

315. $S = 5a^2\sqrt{3}$

316. $V = \dfrac{5a^3\left(3+\sqrt{5}\right)}{12}$

Dodecahedron

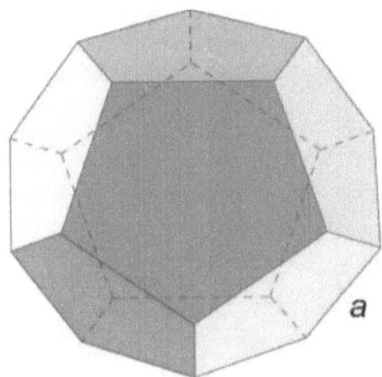

Figure 46.

317. $r = \dfrac{a\sqrt{10(25+11\sqrt{5})}}{2}$

318. $R = \dfrac{a\sqrt{3}(1+\sqrt{5})}{4}$

319. $S = 3a^2\sqrt{5(5+2\sqrt{5})}$

320. $V = \dfrac{a^3(15+7\sqrt{5})}{4}$

3.30 Right Circular Cylinder

Radius of base: R
Diameter of base: d

Height: H
Lateral surface area: S_L
Area of base: S_B
Total surface area: S
Volume: V

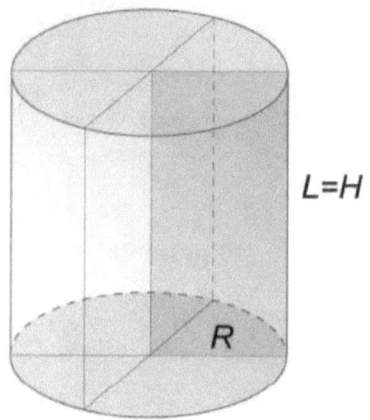

Figure 47.

321. $S_L = 2\pi R H$

322. $S = S_L + 2S_B = 2\pi R(H+R) = \pi d\left(H + \dfrac{d}{2}\right)$

323. $V = S_B H = \pi R^2 H$

3.31 Right Circular Cylinder with an Oblique Plane Face

Radius of base: R
The greatest height of a side: h_1
The shortest height of a side: h_2
Lateral surface area: S_L
Area of plane end faces: S_B
Total surface area: S
Volume: V

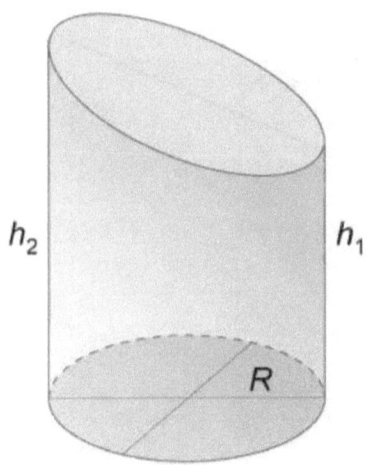

Figure 48.

324. $S_L = \pi R(h_1 + h_2)$

325. $S_B = \pi R^2 + \pi R \sqrt{R^2 + \left(\dfrac{h_1 - h_2}{2}\right)^2}$

326. $S = S_L + S_B = \pi R \left[h_1 + h_2 + R + \sqrt{R^2 + \left(\dfrac{h_1 - h_2}{2}\right)^2} \right]$

327. $V = \dfrac{\pi R^2}{2}(h_1 + h_2)$

3.32 Right Circular Cone

Radius of base: R
Diameter of base: d
Height: H
Slant height: m
Lateral surface area: S_L
Area of base: S_B
Total surface area: S
Volume: V

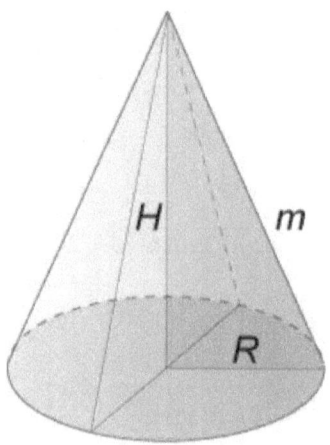

Figure 49.

328. $H = \sqrt{m^2 - R^2}$

329. $S_L = \pi R m = \dfrac{\pi m d}{2}$

330. $S_B = \pi R^2$

331. $S = S_L + S_B = \pi R(m + R) = \dfrac{1}{2}\pi d\left(m + \dfrac{d}{2}\right)$

332. $V = \dfrac{1}{3}S_B H = \dfrac{1}{3}\pi R^2 H$

3.33 Frustum of a Right Circular Cone

Radius of bases: R, r
Height: H
Slant height: m
Scale factor: k
Area of bases: S_1, S_2
Lateral surface area: S_L
Total surface area: S
Volume: V

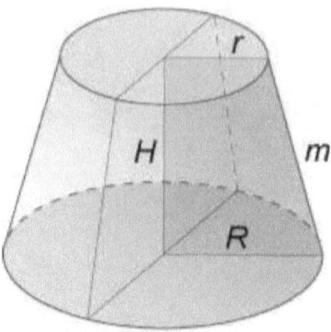

Figure 50.

333. $H = \sqrt{m^2 - (R-r)^2}$

334. $\dfrac{R}{r} = k$

335. $\dfrac{S_2}{S_1} = \dfrac{R^2}{r^2} = k^2$

336. $S_L = \pi m (R + r)$

337. $S = S_1 + S_2 + S_L = \pi \left[R^2 + r^2 + m(R+r) \right]$

338. $V = \dfrac{h}{3} \left(S_1 + \sqrt{S_1 S_2} + S_2 \right)$

339. $V = \dfrac{hS_1}{3} \left[1 + \dfrac{R}{r} + \left(\dfrac{R}{r}\right)^2 \right] = \dfrac{hS_1}{3} \left[1 + k + k^2 \right]$

3.34 Sphere

Radius: R
Diameter: d
Surface area: S
Volume: V

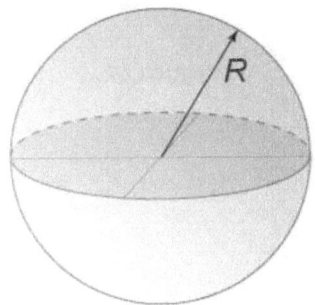

Figure 51.

340. $S = 4\pi R^2$

341. $V = \dfrac{4}{3}\pi R^3 H = \dfrac{1}{6}\pi d^3 = \dfrac{1}{3}SR$

3.35 Spherical Cap

Radius of sphere: R
Radius of base: r
Height: h
Area of plane face: S_B
Area of spherical cap: S_C
Total surface area: S
Volume: V

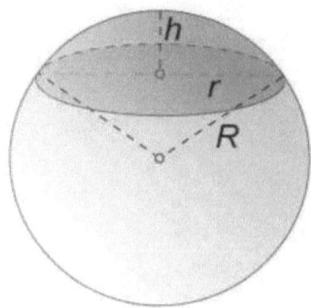

Figure 52.

342. $R = \dfrac{r^2 + h^2}{2h}$

343. $S_B = \pi r^2$

344. $S_C = \pi(h^2 + r^2)$

345. $S = S_B + S_C = \pi(h^2 + 2r^2) = \pi(2Rh + r^2)$

346. $V = \dfrac{\pi}{6}h^2(3R - h) = \dfrac{\pi}{6}h(3r^2 + h^2)$

3.36 Spherical Sector

Radius of sphere: R
Radius of base of spherical cap: r
Height: h
Total surface area: S
Volume: V

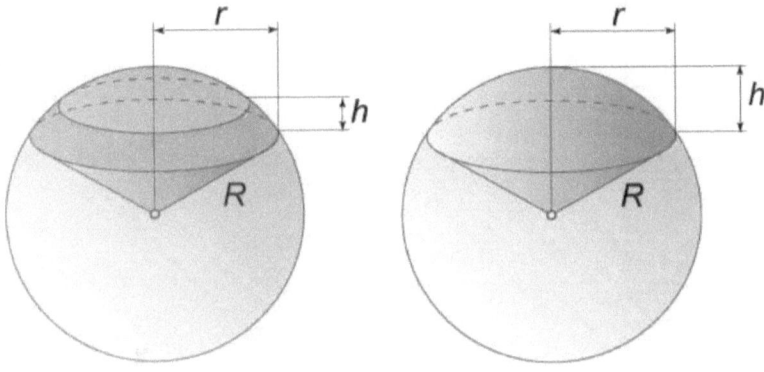

Figure 53.

347. $S = \pi R(2h + r)$

348. $V = \dfrac{2}{3}\pi R^2 h$

Note: The given formulas are correct both for "open" and "closed" spherical sector.

3.37 Spherical Segment

Radius of sphere: R
Radius of bases: r_1, r_2
Height: h
Area of spherical surface: S_S
Area of plane end faces: S_1, S_2
Total surface area: S
Volume: V

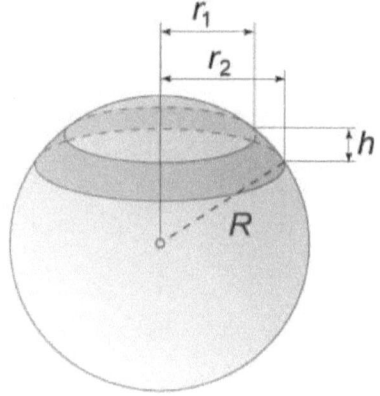

Figure 54.

349. $S_S = 2\pi Rh$

350. $S = S_S + S_1 + S_2 = \pi(2Rh + r_1^2 + r_2^2)$

351. $V = \dfrac{1}{6}\pi h(3r_1^2 + 3r_2^2 + h^2)$

3.38 Spherical Wedge

> Radius: R
> Dihedral angle in degrees: x
> Dihedral angle in radians: α
> Area of spherical lune: S_L
> Total surface area: S
> Volume: V

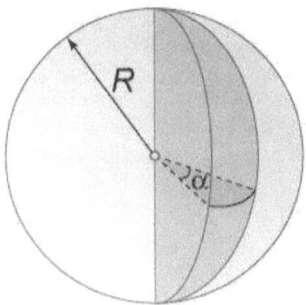

Figure 55.

352. $S_L = \dfrac{\pi R^2}{90}\alpha = 2R^2 x$

353. $S = \pi R^2 + \dfrac{\pi R^2}{90}\alpha = \pi R^2 + 2R^2 x$

354. $V = \dfrac{\pi R^3}{270}\alpha = \dfrac{2}{3}R^3 x$

3.39 Ellipsoid

Semi-axes: a, b, c
Volume: V

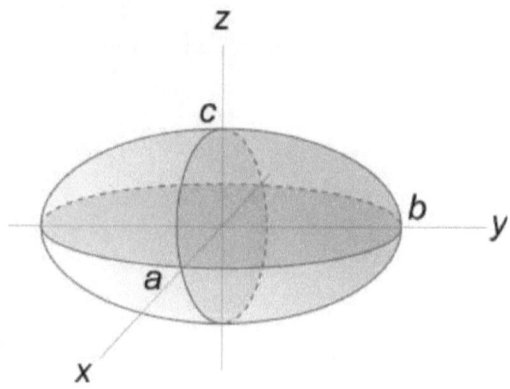

Figure 56.

355. $V = \dfrac{4}{3}\pi abc$

Prolate Spheroid

Semi-axes: a, b, b (a > b)
Surface area: S
Volume: V

356. $S = 2\pi b \left(b + \dfrac{a \arcsin e}{e} \right)$,

where $e = \dfrac{\sqrt{a^2 - b^2}}{a}$.

357. $V = \dfrac{4}{3}\pi b^2 a$

Oblate Spheroid

Semi-axes: a, b, b (a < b)
Surface area: S
Volume: V

358. $S = 2\pi b \left(b + \dfrac{a \operatorname{arcsinh}\left(\dfrac{be}{a}\right)}{be/a} \right)$,

where $e = \dfrac{\sqrt{b^2 - a^2}}{b}$.

359. $V = \dfrac{4}{3}\pi b^2 a$

3.40 Circular Torus

Major radius: R
Minor radius: r
Surface area: S
Volume: V

CHAPTER 3. GEOMETRY

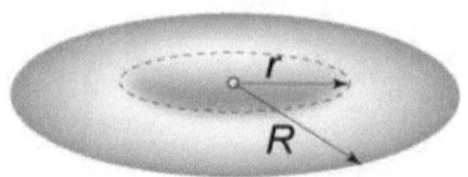

Picture 57.

360. $S = 4\pi^2 Rr$

361. $V = 2\pi^2 Rr^2$

Chapter 4
Trigonometry

Angles: α, β
Real numbers (coordinates of a point): x, y
Whole number: k

4.1 Radian and Degree Measures of Angles

362. $1 \text{ rad} = \dfrac{180°}{\pi} \approx 57°17'45''$

363. $1° = \dfrac{\pi}{180} \text{ rad} \approx 0.017453 \text{ rad}$

364. $1' = \dfrac{\pi}{180 \cdot 60} \text{ rad} \approx 0.000291 \text{ rad}$

365. $1'' = \dfrac{\pi}{180 \cdot 3600} \text{ rad} \approx 0.000005 \text{ rad}$

366.

Angle (degrees)	0	30	45	60	90	180	270	360
Angle (radians)	0	$\dfrac{\pi}{6}$	$\dfrac{\pi}{4}$	$\dfrac{\pi}{3}$	$\dfrac{\pi}{2}$	π	$\dfrac{3\pi}{2}$	2π

4.2 Definitions and Graphs of Trigonometric Functions

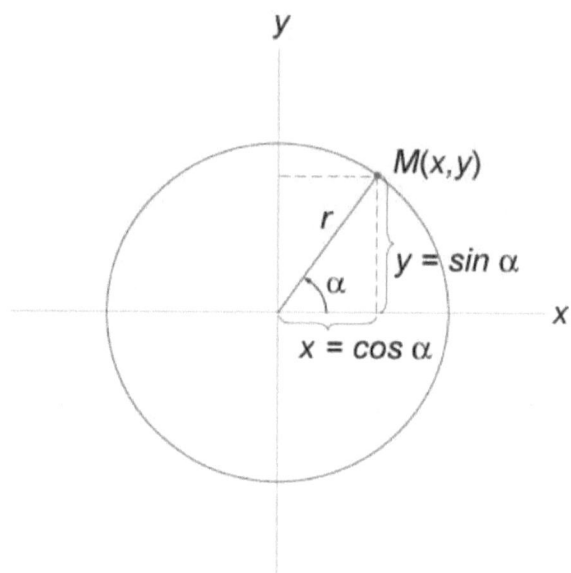

Figure 58.

367. $\sin \alpha = \dfrac{y}{r}$

368. $\cos \alpha = \dfrac{x}{r}$

369. $\tan \alpha = \dfrac{y}{x}$

370. $\cot \alpha = \dfrac{x}{y}$

371. $\sec \alpha = \dfrac{r}{x}$

372. $\operatorname{cosec} \alpha = \dfrac{r}{y}$

373. Sine Function
$y = \sin x$, $-1 \leq \sin x \leq 1$.

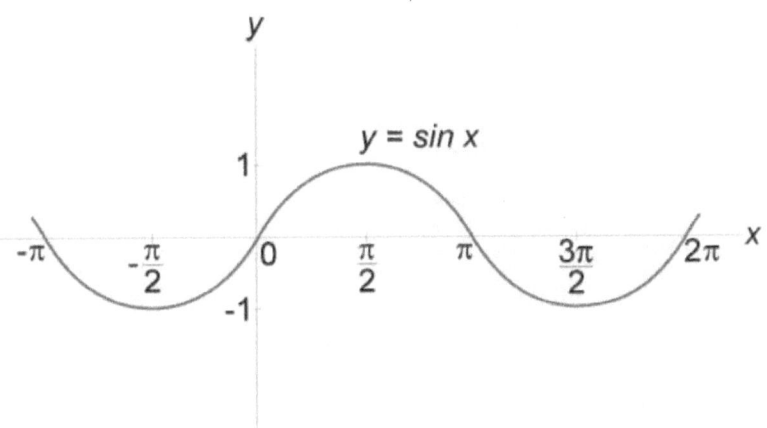

Figure 59.

374. Cosine Function
$y = \cos x$, $-1 \leq \cos x \leq 1$.

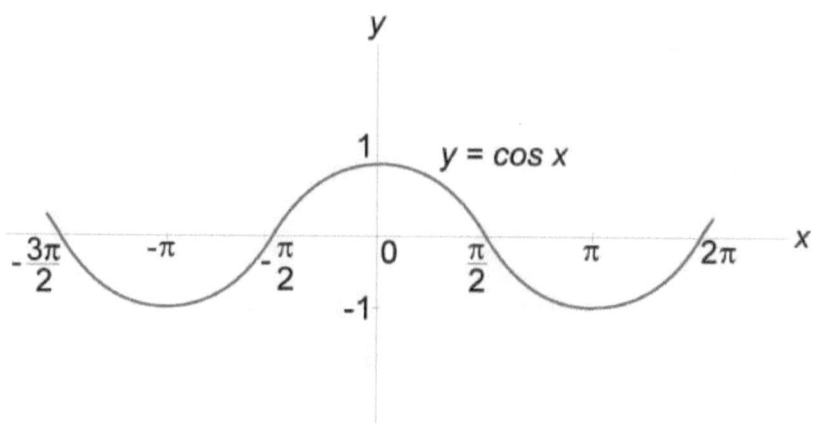

Figure 60.

375. Tangent Function

$$y = \tan x, \quad x \neq (2k+1)\frac{\pi}{2}, \quad -\infty \leq \tan x \leq \infty.$$

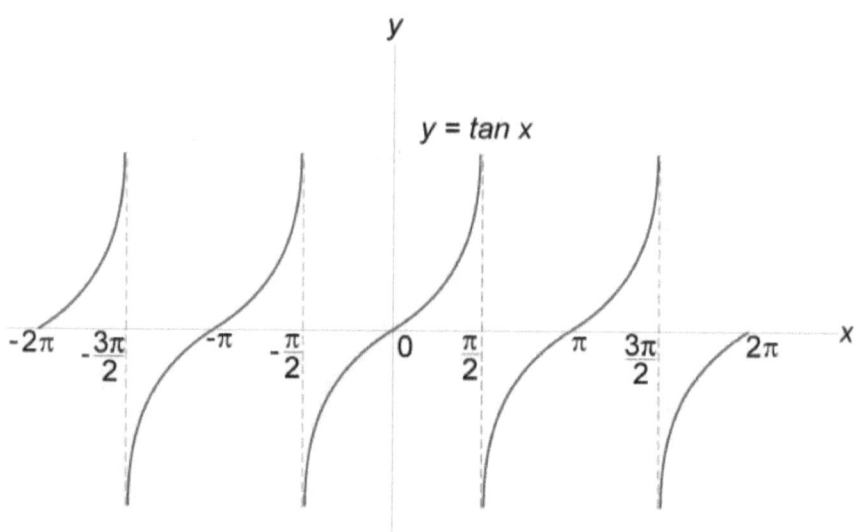

Figure 61.

Cotangent Function
$y = \cot x$, $x \neq k\pi$, $-\infty \leq \cot x \leq \infty$.

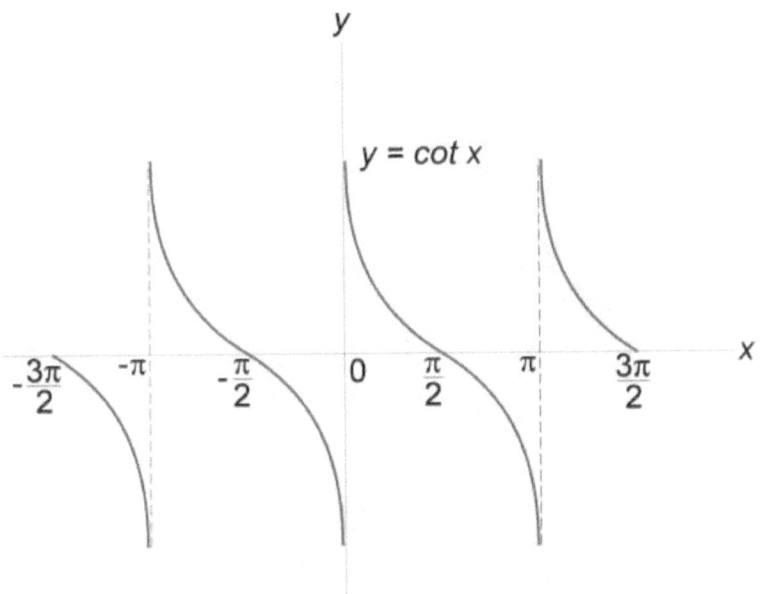

Figure 62.

377. **Secant Function**
$y = \sec x$, $x \neq (2k+1)\dfrac{\pi}{2}$.

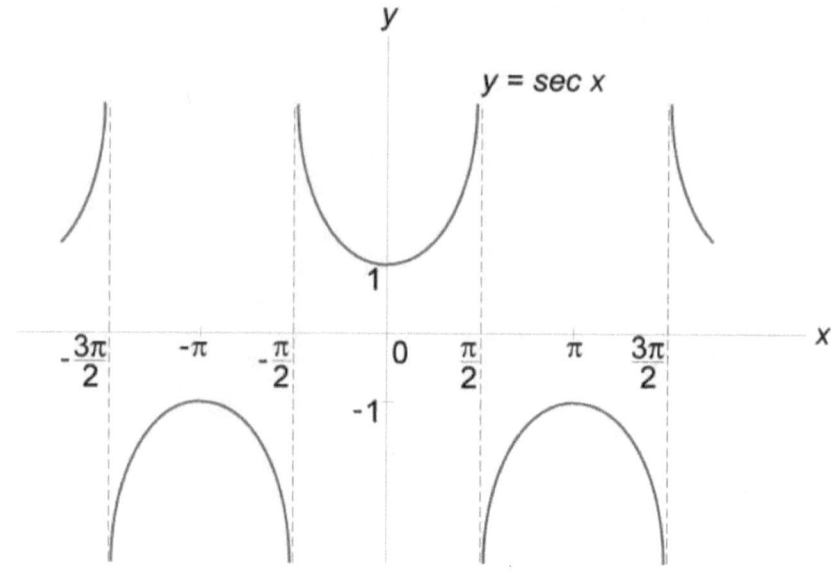

Figure 63.

378. Cosecant Function
$y = \operatorname{cosec} x$, $x \neq k\pi$.

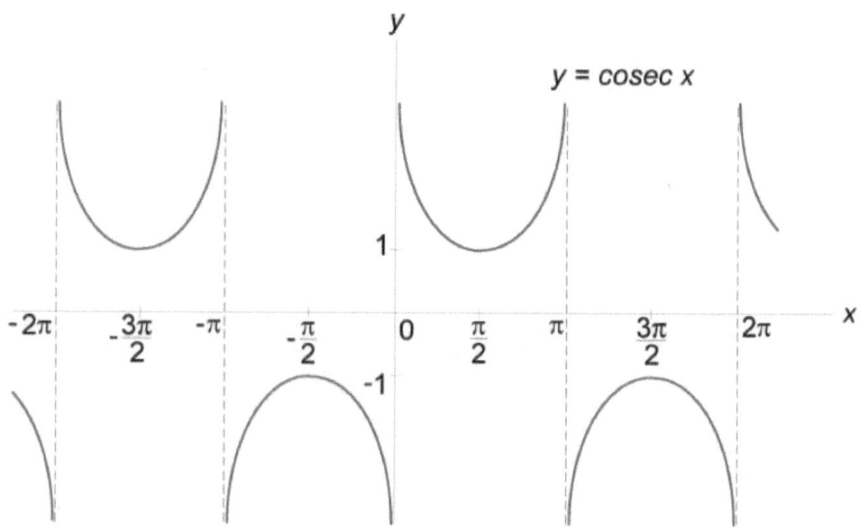

Figure 64.

4.3. Signs of Trigonometric Functions

379.

Quadrant	Sin α	Cos α	Tan α	Cot α	Sec α	Cosec α
I	+	+	+	+	+	+
II	+					+
III			+	+		
IV		+			+	

380.

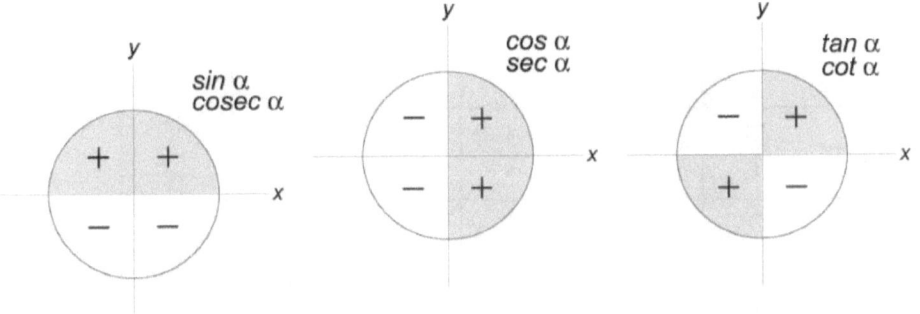

Figure 65.

4.4 Trigonometric Functions of Common Angles

381.

$\alpha°$	α rad	$\sin \alpha$	$\cos \alpha$	$\tan \alpha$	$\cot \alpha$	$\sec \alpha$	$\csc \alpha$
0	0	0	1	0	∞	1	∞
30	$\dfrac{\pi}{6}$	$\dfrac{1}{2}$	$\dfrac{\sqrt{3}}{2}$	$\dfrac{1}{\sqrt{3}}$	$\sqrt{3}$	$\dfrac{2}{\sqrt{3}}$	2
45	$\dfrac{\pi}{4}$	$\dfrac{\sqrt{2}}{2}$	$\dfrac{\sqrt{2}}{2}$	1	1	$\sqrt{2}$	$\sqrt{2}$
60	$\dfrac{\pi}{3}$	$\dfrac{\sqrt{3}}{2}$	$\dfrac{1}{2}$	$\sqrt{3}$	$\dfrac{1}{\sqrt{3}}$	2	$\dfrac{2}{\sqrt{3}}$
90	$\dfrac{\pi}{2}$	1	0	∞	0	∞	1
120	$\dfrac{2\pi}{3}$	$\dfrac{\sqrt{3}}{2}$	$-\dfrac{1}{2}$	$-\sqrt{3}$	$-\dfrac{1}{\sqrt{3}}$	-2	$\dfrac{2}{\sqrt{3}}$
180	π	0	-1	0	∞	-1	∞
270	$\dfrac{3\pi}{2}$	-1	0	∞	0	∞	-1
360	2π	0	1	0	∞	1	∞

382.

$\alpha°$	α rad	$\sin \alpha$	$\cos \alpha$	$\tan \alpha$	$\cot \alpha$
15	$\dfrac{\pi}{12}$	$\dfrac{\sqrt{6}-\sqrt{2}}{4}$	$\dfrac{\sqrt{6}+\sqrt{2}}{4}$	$2-\sqrt{3}$	$2+\sqrt{3}$
18	$\dfrac{\pi}{10}$	$\dfrac{\sqrt{5}-1}{4}$	$\dfrac{\sqrt{10+2\sqrt{5}}}{4}$	$\sqrt{\dfrac{5-2\sqrt{5}}{5}}$	$\sqrt{5+2\sqrt{5}}$
36	$\dfrac{\pi}{5}$	$\dfrac{\sqrt{10-2\sqrt{5}}}{4}$	$\dfrac{\sqrt{5}+1}{4}$	$\dfrac{\sqrt{10-2\sqrt{5}}}{\sqrt{5}+1}$	$\dfrac{\sqrt{5}+1}{\sqrt{10-2\sqrt{5}}}$
54	$\dfrac{3\pi}{10}$	$\dfrac{\sqrt{5}+1}{4}$	$\dfrac{\sqrt{10-2\sqrt{5}}}{4}$	$\dfrac{\sqrt{5}+1}{\sqrt{10-2\sqrt{5}}}$	$\dfrac{\sqrt{10-2\sqrt{5}}}{\sqrt{5}+1}$
72	$\dfrac{2\pi}{5}$	$\dfrac{\sqrt{10+2\sqrt{5}}}{4}$	$\dfrac{\sqrt{5}-1}{4}$	$\sqrt{5+2\sqrt{5}}$	$\sqrt{\dfrac{5-2\sqrt{5}}{5}}$
75	$\dfrac{5\pi}{12}$	$\dfrac{\sqrt{6}+\sqrt{2}}{4}$	$\dfrac{\sqrt{6}-\sqrt{2}}{4}$	$2+\sqrt{3}$	$2-\sqrt{3}$

4.5 Most Important Formulas

383. $\sin^2 \alpha + \cos^2 \alpha = 1$

384. $\sec^2 \alpha - \tan^2 \alpha = 1$

385. $\csc^2 \alpha - \cot^2 \alpha = 1$

386. $\tan \alpha = \dfrac{\sin \alpha}{\cos \alpha}$

387. $\cot\alpha = \dfrac{\cos\alpha}{\sin\alpha}$

388. $\tan\alpha \cdot \cot\alpha = 1$

389. $\sec\alpha = \dfrac{1}{\cos\alpha}$

390. $\operatorname{cosec}\alpha = \dfrac{1}{\sin\alpha}$

4.6 Reduction Formulas

391.

β	$\sin\beta$	$\cos\beta$	$\tan\beta$	$\cot\beta$
$-\alpha$	$-\sin\alpha$	$+\cos\alpha$	$-\tan\alpha$	$-\cot\alpha$
$90°-\alpha$	$+\cos\alpha$	$+\sin\alpha$	$+\cot\alpha$	$+\tan\alpha$
$90°+\alpha$	$+\cos\alpha$	$-\sin\alpha$	$-\cot\alpha$	$-\tan\alpha$
$180°-\alpha$	$+\sin\alpha$	$-\cos\alpha$	$-\tan\alpha$	$-\cot\alpha$
$180°+\alpha$	$-\sin\alpha$	$-\cos\alpha$	$+\tan\alpha$	$+\cot\alpha$
$270°-\alpha$	$-\cos\alpha$	$-\sin\alpha$	$+\cot\alpha$	$+\tan\alpha$
$270°+\alpha$	$-\cos\alpha$	$+\sin\alpha$	$-\cot\alpha$	$-\tan\alpha$
$360°-\alpha$	$-\sin\alpha$	$+\cos\alpha$	$-\tan\alpha$	$-\cot\alpha$
$360°+\alpha$	$+\sin\alpha$	$+\cos\alpha$	$+\tan\alpha$	$+\cot\alpha$

4.7 Periodicity of Trigonometric Functions

392. $\sin(\alpha \pm 2\pi n) = \sin \alpha$, period 2π or $360°$.

393. $\cos(\alpha \pm 2\pi n) = \cos \alpha$, period 2π or $360°$.

394. $\tan(\alpha \pm \pi n) = \tan \alpha$, period π or $180°$.

395. $\cot(\alpha \pm \pi n) = \cot \alpha$, period π or $180°$.

4.8 Relations between Trigonometric Functions

396. $\sin \alpha = \pm\sqrt{1 - \cos^2 \alpha} = \pm\sqrt{\frac{1}{2}(1 - \cos 2\alpha)} = 2\cos^2\left(\frac{\alpha}{2} - \frac{\pi}{4}\right) - 1$

$$= \frac{2\tan\frac{\alpha}{2}}{1 + \tan^2\frac{\alpha}{2}}$$

397. $\cos \alpha = \pm\sqrt{1 - \sin^2 \alpha} = \pm\sqrt{\frac{1}{2}(1 + \cos 2\alpha)} = 2\cos^2\frac{\alpha}{2} - 1$

$$= \frac{1 - \tan^2\frac{\alpha}{2}}{1 + \tan^2\frac{\alpha}{2}}$$

398. $\tan \alpha = \dfrac{\sin \alpha}{\cos \alpha} = \pm\sqrt{\sec^2 \alpha - 1} = \dfrac{\sin 2\alpha}{1 + \cos 2\alpha} = \dfrac{1 - \cos 2\alpha}{\sin 2\alpha}$

$$= \pm\sqrt{\frac{1-\cos 2\alpha}{1+\cos 2\alpha}} = \frac{2\tan\frac{\alpha}{2}}{1+\tan^2\frac{\alpha}{2}}$$

399. $\cot\alpha = \dfrac{\cos\alpha}{\sin\alpha} = \pm\sqrt{\csc^2\alpha - 1} = \dfrac{1+\cos 2\alpha}{\sin 2\alpha} = \dfrac{\sin 2\alpha}{1-\cos 2\alpha}$

$$= \pm\sqrt{\frac{1+\cos 2\alpha}{1-\cos 2\alpha}} = \frac{1-\tan^2\frac{\alpha}{2}}{2\tan\frac{\alpha}{2}}$$

400. $\sec\alpha = \dfrac{1}{\cos\alpha} = \pm\sqrt{1+\tan^2\alpha} = \dfrac{1+\tan^2\frac{\alpha}{2}}{1-\tan^2\frac{\alpha}{2}}$

401. $\csc\alpha = \dfrac{1}{\sin\alpha} = \pm\sqrt{1+\cot^2\alpha} = \dfrac{1+\tan^2\frac{\alpha}{2}}{2\tan\frac{\alpha}{2}}$

4.9 Addition and Subtraction Formulas

402. $\sin(\alpha+\beta) = \sin\alpha\cos\beta + \sin\beta\cos\alpha$

403. $\sin(\alpha-y) = \sin\alpha\cos\beta - \sin\beta\cos\alpha$

404. $\cos(\alpha+\beta) = \cos\alpha\cos\beta - \sin\alpha\sin\beta$

405. $\cos(\alpha-\beta) = \cos\alpha\cos\beta + \sin\alpha\sin\beta$

406. $\tan(\alpha+\beta) = \dfrac{\tan\alpha + \tan\beta}{1 - \tan\alpha\tan\beta}$

407. $\tan(\alpha-\beta) = \dfrac{\tan\alpha - \tan\beta}{1 + \tan\alpha\tan\beta}$

408. $\cot(\alpha+\beta) = \dfrac{1 - \tan\alpha\tan\beta}{\tan\alpha + \tan\beta}$

409. $\cot(\alpha-\beta) = \dfrac{1 + \tan\alpha\tan\beta}{\tan\alpha - \tan\beta}$

4.10 Double Angle Formulas

410. $\sin 2\alpha = 2\sin\alpha \cdot \cos\alpha$

411. $\cos 2\alpha = \cos^2\alpha - \sin^2\alpha = 1 - 2\sin^2\alpha = 2\cos^2\alpha - 1$

412. $\tan 2\alpha = \dfrac{2\tan\alpha}{1 - \tan^2\alpha} = \dfrac{2}{\cot\alpha - \tan\alpha}$

413. $\cot 2\alpha = \dfrac{\cot^2\alpha - 1}{2\cot\alpha} = \dfrac{\cot\alpha - \tan\alpha}{2}$

4.11 Multiple Angle Formulas

414. $\sin 3\alpha = 3\sin\alpha - 4\sin^3\alpha = 3\cos^2\alpha \cdot \sin\alpha - \sin^3\alpha$

415. $\sin 4\alpha = 4\sin\alpha \cdot \cos\alpha - 8\sin^3\alpha \cdot \cos\alpha$

416. $\sin 5\alpha = 5\sin\alpha - 20\sin^3\alpha + 16\sin^5\alpha$

417. $\cos 3\alpha = 4\cos^3\alpha - 3\cos\alpha = \cos^3\alpha - 3\cos\alpha \cdot \sin^2\alpha$

418. $\cos 4\alpha = 8\cos^4\alpha - 8\cos^2\alpha + 1$

419. $\cos 5\alpha = 16\cos^5\alpha - 20\cos^3\alpha + 5\cos\alpha$

420. $\tan 3\alpha = \dfrac{3\tan\alpha - \tan^3\alpha}{1 - 3\tan^2\alpha}$

421. $\tan 4\alpha = \dfrac{4\tan\alpha - 4\tan^3\alpha}{1 - 6\tan^2\alpha + \tan^4\alpha}$

422. $\tan 5\alpha = \dfrac{\tan^5\alpha - 10\tan^3\alpha + 5\tan\alpha}{1 - 10\tan^2\alpha + 5\tan^4\alpha}$

423. $\cot 3\alpha = \dfrac{\cot^3\alpha - 3\cot\alpha}{3\cot^2\alpha - 1}$

424. $\cot 4\alpha = \dfrac{1 - 6\tan^2\alpha + \tan^4\alpha}{4\tan\alpha - 4\tan^3\alpha}$

425. $\cot 5\alpha = \dfrac{1 - 10\tan^2\alpha + 5\tan^4\alpha}{\tan^5\alpha - 10\tan^3\alpha + 5\tan\alpha}$

4.12 Half Angle Formulas

426. $\sin\dfrac{\alpha}{2} = \pm\sqrt{\dfrac{1-\cos\alpha}{2}}$

427. $\cos\dfrac{\alpha}{2} = \pm\sqrt{\dfrac{1+\cos\alpha}{2}}$

428. $\tan\dfrac{\alpha}{2} = \pm\sqrt{\dfrac{1-\cos\alpha}{1+\cos\alpha}} = \dfrac{\sin\alpha}{1+\cos\alpha} = \dfrac{1-\cos\alpha}{\sin\alpha} = \csc\alpha - \cot\alpha$

429. $\cot\dfrac{\alpha}{2} = \pm\sqrt{\dfrac{1+\cos\alpha}{1-\cos\alpha}} = \dfrac{\sin\alpha}{1-\cos\alpha} = \dfrac{1+\cos\alpha}{\sin\alpha} = \csc\alpha + \cot\alpha$

4.13 Half Angle Tangent Identities

430. $\sin\alpha = \dfrac{2\tan\dfrac{\alpha}{2}}{1+\tan^2\dfrac{\alpha}{2}}$

431. $\cos\alpha = \dfrac{1-\tan^2\dfrac{\alpha}{2}}{1+\tan^2\dfrac{\alpha}{2}}$

432. $\tan\alpha = \dfrac{2\tan\dfrac{\alpha}{2}}{1-\tan^2\dfrac{\alpha}{2}}$

433. $\cot\alpha = \dfrac{1-\tan^2\dfrac{\alpha}{2}}{2\tan\dfrac{\alpha}{2}}$

4.14 Transforming of Trigonometric Expressions to Product

434. $\sin\alpha + \sin\beta = 2\sin\dfrac{\alpha+\beta}{2}\cos\dfrac{\alpha-\beta}{2}$

435. $\sin\alpha - \sin\beta = 2\cos\dfrac{\alpha+\beta}{2}\sin\dfrac{\alpha-\beta}{2}$

436. $\cos\alpha + \cos\beta = 2\cos\dfrac{\alpha+\beta}{2}\cos\dfrac{\alpha-\beta}{2}$

437. $\cos\alpha - \cos\beta = -2\sin\dfrac{\alpha+\beta}{2}\sin\dfrac{\alpha-\beta}{2}$

CHAPTER 4. TRIGONOMETRY

438. $\tan\alpha + \tan\beta = \dfrac{\sin(\alpha+\beta)}{\cos\alpha\cdot\cos\beta}$

439. $\tan\alpha - \tan\beta = \dfrac{\sin(\alpha-\beta)}{\cos\alpha\cdot\cos\beta}$

440. $\cot\alpha + \cot\beta = \dfrac{\sin(\beta+\alpha)}{\sin\alpha\cdot\sin\beta}$

441. $\cot\alpha - \cot\beta = \dfrac{\sin(\beta-\alpha)}{\sin\alpha\cdot\sin\beta}$

442. $\cos\alpha + \sin\alpha = \sqrt{2}\cos\left(\dfrac{\pi}{4}-\alpha\right) = \sqrt{2}\sin\left(\dfrac{\pi}{4}+\alpha\right)$

443. $\cos\alpha - \sin\alpha = \sqrt{2}\sin\left(\dfrac{\pi}{4}-\alpha\right) = \sqrt{2}\cos\left(\dfrac{\pi}{4}+\alpha\right)$

444. $\tan\alpha + \cot\beta = \dfrac{\cos(\alpha-\beta)}{\cos\alpha\cdot\sin\beta}$

445. $\tan\alpha - \cot\beta = -\dfrac{\cos(\alpha+\beta)}{\cos\alpha\cdot\sin\beta}$

446. $1 + \cos\alpha = 2\cos^2\dfrac{\alpha}{2}$

447. $1 - \cos\alpha = 2\sin^2\dfrac{\alpha}{2}$

448. $1+\sin\alpha = 2\cos^2\left(\dfrac{\pi}{4}-\dfrac{\alpha}{2}\right)$

449. $1-\sin\alpha = 2\sin^2\left(\dfrac{\pi}{4}-\dfrac{\alpha}{2}\right)$

4.15 Transforming of Trigonometric Expressions to Sum

450. $\sin\alpha\cdot\sin\beta = \dfrac{\cos(\alpha-\beta)-\cos(\alpha+\beta)}{2}$

451. $\cos\alpha\cdot\cos\beta = \dfrac{\cos(\alpha-\beta)+\cos(\alpha+\beta)}{2}$

452. $\sin\alpha\cdot\cos\beta = \dfrac{\sin(\alpha-\beta)+\sin(\alpha+\beta)}{2}$

453. $\tan\alpha\cdot\tan\beta = \dfrac{\tan\alpha+\tan\beta}{\cot\alpha+\cot\beta}$

454. $\cot\alpha\cdot\cot\beta = \dfrac{\cot\alpha+\cot\beta}{\tan\alpha+\tan\beta}$

455. $\tan\alpha\cdot\cot\beta = \dfrac{\tan\alpha+\cot\beta}{\cot\alpha+\tan\beta}$

4.16 Powers of Trigonometric Functions

456. $\sin^2 \alpha = \dfrac{1-\cos 2\alpha}{2}$

457. $\sin^3 \alpha = \dfrac{3\sin \alpha - \sin 3\alpha}{4}$

458. $\sin^4 \alpha = \dfrac{\cos 4\alpha - 4\cos 2\alpha + 3}{8}$

459. $\sin^5 \alpha = \dfrac{10\sin \alpha - 5\sin 3\alpha + \sin 5\alpha}{16}$

460. $\sin^6 \alpha = \dfrac{10 - 15\cos 2\alpha + 6\cos 4\alpha - \cos 6\alpha}{32}$

461. $\cos^2 \alpha = \dfrac{1+\cos 2\alpha}{2}$

462. $\cos^3 \alpha = \dfrac{3\cos \alpha + \cos 3\alpha}{4}$

463. $\cos^4 \alpha = \dfrac{\cos 4\alpha + 4\cos 2\alpha + 3}{8}$

464. $\cos^5 \alpha = \dfrac{10\cos \alpha + 5\sin 3\alpha + \cos 5\alpha}{16}$

465. $\cos^6 \alpha = \dfrac{10 + 15\cos 2\alpha + 6\cos 4\alpha + \cos 6\alpha}{32}$

4.17 Graphs of Inverse Trigonometric Functions

466. Inverse Sine Function

$$y = \arcsin x, \quad -1 \leq x \leq 1, \quad -\frac{\pi}{2} \leq \arcsin x \leq \frac{\pi}{2}.$$

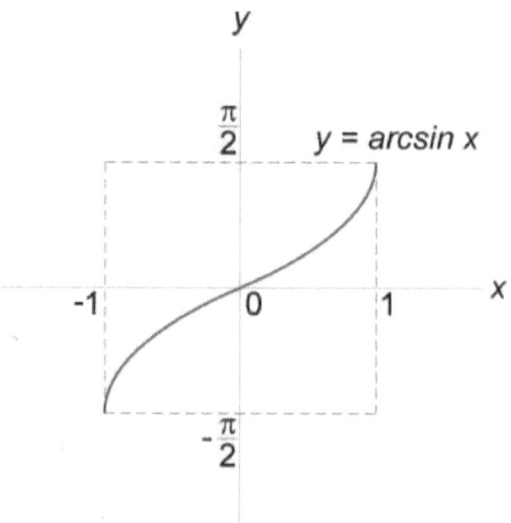

Figure 66.

467. Inverse Cosine Function

$$y = \arccos x, \quad -1 \leq x \leq 1, \quad 0 \leq \arccos x \leq \pi.$$

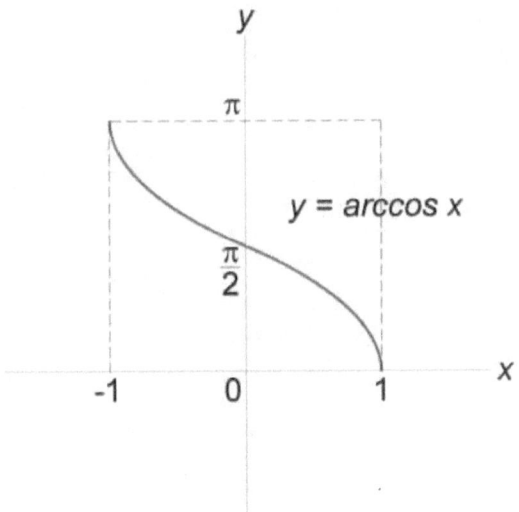

Figure 67.

468. Inverse Tangent Function

$$y = \arctan x, \quad -\infty \leq x \leq \infty, \quad -\frac{\pi}{2} < \arctan x < \frac{\pi}{2}.$$

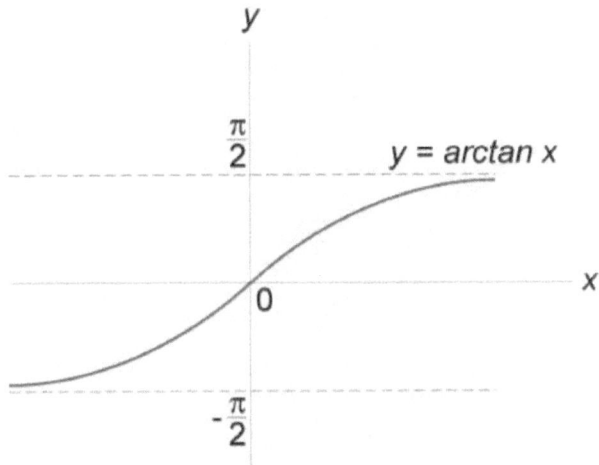

Figure 68.

Inverse Cotangent Function
$y = \text{arccot}\, x$, $-\infty \leq x \leq \infty$, $0 < \text{arccot}\, x < \pi$.

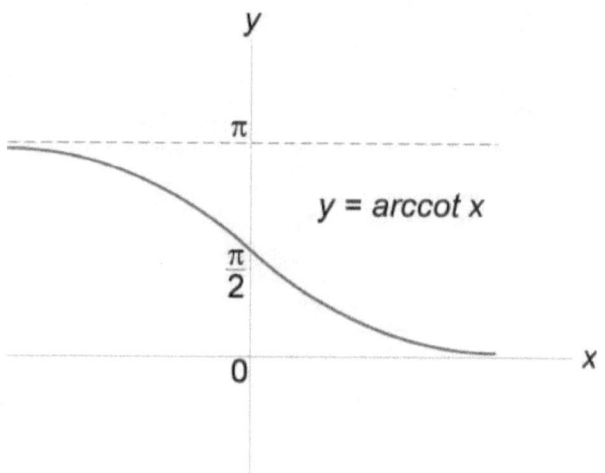

Figure 69.

470. Inverse Secant Function

$y = \text{arcsec}\, x$, $x \in (-\infty, -1] \cup [1, \infty)$, $\text{arcsec}\, x \in \left[0, \dfrac{\pi}{2}\right) \cup \left(\dfrac{\pi}{2}, \pi\right]$.

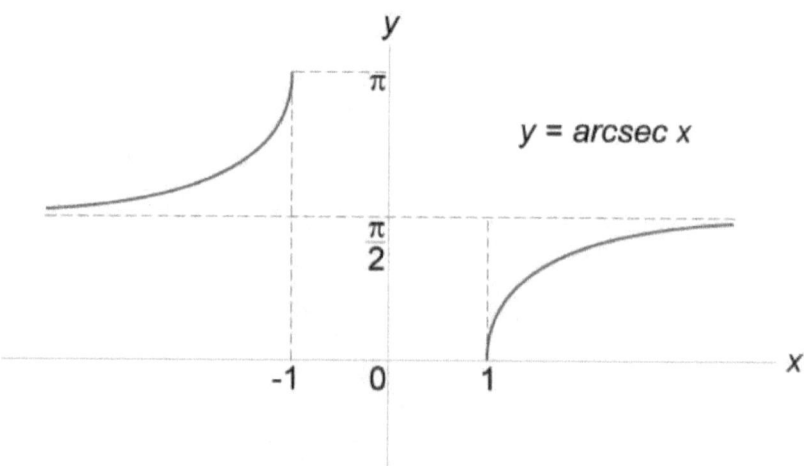

Figure 70.

471. Inverse Cosecant Function

$y = \text{arccsc } x, \ x \in (-\infty, -1] \cup [1, \infty), \ \text{arccsc } x \in \left[-\dfrac{\pi}{2}, 0\right) \cup \left(0, \dfrac{\pi}{2}\right].$

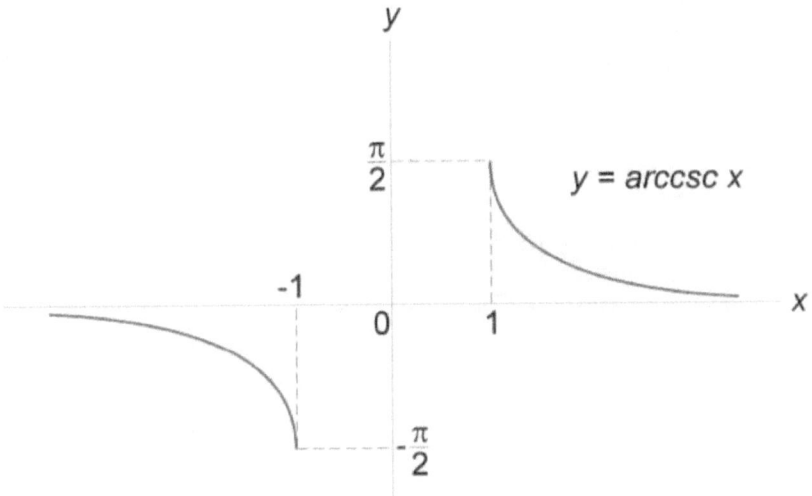

Figure 71.

4.18 Principal Values of Inverse Trigonometric Functions

472.

x	0	$\dfrac{1}{2}$	$\dfrac{\sqrt{2}}{2}$	$\dfrac{\sqrt{3}}{2}$	1
arcsin x	0°	30°	45°	60°	90°
arccos x	90°	60°	45°	30°	0°

x	$-\dfrac{1}{2}$	$-\dfrac{\sqrt{2}}{2}$	$-\dfrac{\sqrt{3}}{2}$	-1	
arcsin x	$-30°$	$-45°$	$-60°$	$-90°$	
arccos x	120°	135°	150°	180°	

473.

x	0	$\dfrac{\sqrt{3}}{3}$	1	$\sqrt{3}$	$-\dfrac{\sqrt{3}}{3}$	-1	$-\sqrt{3}$
arctan x	0°	30°	45°	60°	−30°	−45°	−60°
arccot x	90°	60°	45°	30°	120°	135°	150°

4.19 Relations between Inverse Trigonometric Functions

474. $\arcsin(-x) = -\arcsin x$

475. $\arcsin x = \dfrac{\pi}{2} - \arccos x$

476. $\arcsin x = \arccos\sqrt{1-x^2},\ 0 \leq x \leq 1$.

477. $\arcsin x = -\arccos\sqrt{1-x^2},\ -1 \leq x \leq 0$.

478. $\arcsin x = \arctan\dfrac{x}{\sqrt{1-x^2}},\ x^2 < 1$.

479. $\arcsin x = \mathrm{arccot}\dfrac{\sqrt{1-x^2}}{x},\ 0 < x \leq 1$.

480. $\arcsin x = \mathrm{arccot}\dfrac{\sqrt{1-x^2}}{x} - \pi,\ -1 \leq x < 0$.

481. $\arccos(-x) = \pi - \arccos x$

CHAPTER 4. TRIGONOMETRY

482. $\arccos x = \dfrac{\pi}{2} - \arcsin x$

483. $\arccos x = \arcsin\sqrt{1-x^2}$, $0 \le x \le 1$.

484. $\arccos x = \pi - \arcsin\sqrt{1-x^2}$, $-1 \le x \le 0$.

485. $\arccos x = \arctan\dfrac{\sqrt{1-x^2}}{x}$, $0 < x \le 1$.

486. $\arccos x = \pi + \arctan\dfrac{\sqrt{1-x^2}}{x}$, $-1 \le x < 0$.

487. $\arccos x = \operatorname{arccot}\dfrac{x}{\sqrt{1-x^2}}$, $-1 \le x \le 1$.

488. $\arctan(-x) = -\arctan x$

489. $\arctan x = \dfrac{\pi}{2} - \operatorname{arccot} x$

490. $\arctan x = \arcsin\dfrac{x}{\sqrt{1+x^2}}$

491. $\arctan x = \arccos\dfrac{1}{\sqrt{1+x^2}}$, $x \ge 0$.

492. $\arctan x = -\arccos\dfrac{1}{\sqrt{1+x^2}}$, $x \le 0$.

CHAPTER 4. TRIGONOMETRY

493. $\arctan x = \dfrac{\pi}{2} - \arctan \dfrac{1}{x}$, $x > 0$.

494. $\arctan x = -\dfrac{\pi}{2} - \arctan \dfrac{1}{x}$, $x < 0$.

495. $\arctan x = \operatorname{arccot} \dfrac{1}{x}$, $x > 0$.

496. $\arctan x = \operatorname{arccot} \dfrac{1}{x} - \pi$, $x < 0$.

497. $\operatorname{arccot}(-x) = \pi - \operatorname{arccot} x$

498. $\operatorname{arccot} x = \dfrac{\pi}{2} - \arctan x$

499. $\operatorname{arccot} x = \arcsin \dfrac{1}{\sqrt{1+x^2}}$, $x > 0$.

500. $\operatorname{arccot} x = \pi - \arcsin \dfrac{1}{\sqrt{1+x^2}}$, $x < 0$.

501. $\operatorname{arccot} x = \arccos \dfrac{x}{\sqrt{1+x^2}}$

502. $\operatorname{arccot} x = \arctan \dfrac{1}{x}$, $x > 0$.

503. $\operatorname{arccot} x = \pi + \arctan \dfrac{1}{x}$, $x < 0$.

4.20 Trigonometric Equations

Whole number: n

504. $\sin x = a$, $x = (-1)^n \arcsin a + \pi n$

505. $\cos x = a$, $x = \pm \arccos a + 2\pi n$

506. $\tan x = a$, $x = \arctan a + \pi n$

507. $\cot x = a$, $x = \text{arc cot } a + \pi n$

4.21 Relations to Hyperbolic Functions

Imaginary unit: i

508. $\sin(ix) = i \sinh x$

509. $\tan(ix) = i \tanh x$

510. $\cot(ix) = -i \coth x$

511. $\sec(ix) = \text{sech } x$

512. $\csc(ix) = -i \operatorname{csch} x$

Chapter 5
Matrices and Determinants

Matrices: A, B, C
Elements of a matrix: a_i, b_i, a_{ij}, b_{ij}, c_{ij}
Determinant of a matrix: det A
Minor of an element a_{ij} : M_{ij}
Cofactor of an element a_{ij} : C_{ij}

Transpose of a matrix: A^T, $\tilde{A}$
Adjoint of a matrix: adj A
Trace of a matrix: tr A

Inverse of a matrix: A^{-1}
Real number: k
Real variables: x_i
Natural numbers: m, n

5.1 Determinants

513. Second Order Determinant
$$\det A = \begin{vmatrix} a_1 & b_1 \\ a_2 & b_2 \end{vmatrix} = a_1 b_2 - a_2 b_1$$

514. Third Order Determinant

$$\det A = \begin{vmatrix} a_{11} & a_{12} & a_{13} \\ a_{21} & a_{22} & a_{23} \\ a_{31} & a_{32} & a_{33} \end{vmatrix} = a_{11}a_{22}a_{33} + a_{12}a_{23}a_{31} + a_{13}a_{21}a_{32} - a_{11}a_{23}a_{32} - a_{12}a_{21}a_{33} - a_{13}a_{22}a_{31}$$

515. Sarrus Rule (Arrow Rule)

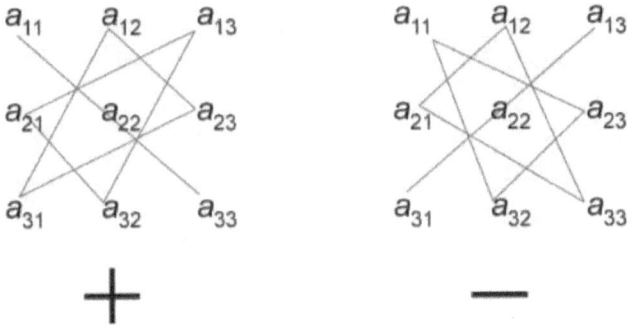

Figure 72.

516. N-th Order Determinant

$$\det A = \begin{vmatrix} a_{11} & a_{12} & \cdots & a_{1j} & \cdots & a_{1n} \\ a_{21} & a_{22} & \cdots & a_{2j} & \cdots & a_{2n} \\ \cdots & \cdots & \cdots & \cdots & \cdots & \cdots \\ a_{i1} & a_{i2} & \cdots & a_{ij} & \cdots & a_{in} \\ \cdots & \cdots & \cdots & \cdots & \cdots & \cdots \\ a_{n1} & a_{n2} & \cdots & a_{nj} & \cdots & a_{nn} \end{vmatrix}$$

517. Minor

The minor M_{ij} associated with the element a_{ij} of n-th order matrix A is the $(n-1)$-th order determinant derived from the matrix A by deletion of its i-th row and j-th column.

518. Cofactor

$$C_{ij} = (-1)^{i+j} M_{ij}$$

519. Laplace Expansion of n-th Order Determinant

Laplace expansion by elements of the i-th row

$$\det A = \sum_{j=1}^{n} a_{ij} C_{ij}, \quad i = 1, 2, \ldots, n.$$

Laplace expansion by elements of the j-th column

$$\det A = \sum_{i=1}^{n} a_{ij} C_{ij}, \quad j = 1, 2, \ldots, n.$$

5.2 Properties of Determinants

520. The value of a determinant remains unchanged if rows are changed to columns and columns to rows.

$$\begin{vmatrix} a_1 & a_2 \\ b_1 & b_2 \end{vmatrix} = \begin{vmatrix} a_1 & b_1 \\ a_2 & b_2 \end{vmatrix}$$

521. If two rows (or two columns) are interchanged, the sign of the determinant is changed.

$$\begin{vmatrix} a_1 & b_1 \\ a_2 & b_2 \end{vmatrix} = -\begin{vmatrix} a_2 & b_2 \\ a_1 & b_1 \end{vmatrix}$$

522. If two rows (or two columns) are identical, the value of the determinant is zero.

$$\begin{vmatrix} a_1 & a_1 \\ a_2 & a_2 \end{vmatrix} = 0$$

523. If the elements of any row (or column) are multiplied by a common factor, the determinant is multiplied by that factor.
$$\begin{vmatrix} ka_1 & kb_1 \\ a_2 & b_2 \end{vmatrix} = k \begin{vmatrix} a_1 & b_1 \\ a_2 & b_2 \end{vmatrix}$$

524. If the elements of any row (or column) are increased (or decreased) by equal multiples of the corresponding elements of any other row (or column), the value of the determinant is unchanged.
$$\begin{vmatrix} a_1 + kb_1 & b_1 \\ a_2 + kb_2 & b_2 \end{vmatrix} = \begin{vmatrix} a_1 & b_1 \\ a_2 & b_2 \end{vmatrix}$$

5.3 Matrices

525. Definition
An $m \times n$ matrix A is a rectangular array of elements (numbers or functions) with m rows and n columns.
$$A = [a_{ij}] = \begin{bmatrix} a_{11} & a_{12} & \cdots & a_{1n} \\ a_{21} & a_{22} & \cdots & a_{2n} \\ \vdots & \vdots & & \vdots \\ a_{m1} & a_{m2} & \cdots & a_{mn} \end{bmatrix}$$

526. Square matrix is a matrix of order $n \times n$.

527. A square matrix $[a_{ij}]$ is symmetric if $a_{ij} = a_{ji}$, i.e. it is symmetric about the leading diagonal.

528. A square matrix $[a_{ij}]$ is skew-symmetric if $a_{ij} = -a_{ji}$.

529. Diagonal matrix is a square matrix with all elements zero except those on the leading diagonal.

530. Unit matrix is a diagonal matrix in which the elements on the leading diagonal are all unity. The unit matrix is denoted by I.

531. A null matrix is one whose elements are all zero.

5.4 Operations with Matrices

532. Two matrices A and B are equal if, and only if, they are both of the same shape $m \times n$ and corresponding elements are equal.

533. Two matrices A and B can be added (or subtracted) of, and only if, they have the same shape $m \times n$. If

$$A = [a_{ij}] = \begin{bmatrix} a_{11} & a_{12} & \cdots & a_{1n} \\ a_{21} & a_{22} & \cdots & a_{2n} \\ \vdots & \vdots & & \vdots \\ a_{m1} & a_{m2} & \cdots & a_{mn} \end{bmatrix},$$

$$B = [b_{ij}] = \begin{bmatrix} b_{11} & b_{12} & \cdots & b_{1n} \\ b_{21} & b_{22} & \cdots & b_{2n} \\ \vdots & \vdots & & \vdots \\ b_{m1} & b_{m2} & \cdots & b_{mn} \end{bmatrix},$$

then
$$A+B = \begin{bmatrix} a_{11}+b_{11} & a_{12}+b_{12} & \cdots & a_{1n}+b_{1n} \\ a_{21}+b_{21} & a_{22}+b_{22} & \cdots & a_{2n}+b_{2n} \\ \vdots & \vdots & & \vdots \\ a_{m1}+b_{m1} & a_{m2}+b_{m2} & \cdots & a_{mn}+b_{mn} \end{bmatrix}.$$

534. If k is a scalar, and $A = [a_{ij}]$ is a matrix, then
$$kA = [ka_{ij}] = \begin{bmatrix} ka_{11} & ka_{12} & \cdots & ka_{1n} \\ ka_{21} & ka_{22} & \cdots & ka_{2n} \\ \vdots & \vdots & & \vdots \\ ka_{m1} & ka_{m2} & \cdots & ka_{mn} \end{bmatrix}.$$

535. Multiplication of Two Matrices
Two matrices can be multiplied together only when the number of columns in the first is equal to the number of rows in the second.

If
$$A = [a_{ij}] = \begin{bmatrix} a_{11} & a_{12} & \cdots & a_{1n} \\ a_{21} & a_{22} & \cdots & a_{2n} \\ \vdots & \vdots & & \vdots \\ a_{m1} & a_{m2} & \cdots & a_{mn} \end{bmatrix},$$

$$B = [b_{ij}] = \begin{bmatrix} b_{11} & b_{12} & \cdots & b_{1k} \\ b_{21} & b_{22} & \cdots & b_{2k} \\ \vdots & \vdots & & \vdots \\ b_{n1} & b_{n2} & \cdots & b_{nk} \end{bmatrix},$$

then
$$AB = C = \begin{bmatrix} c_{11} & c_{12} & \cdots & c_{1k} \\ c_{21} & c_{22} & \cdots & c_{2k} \\ \vdots & \vdots & & \vdots \\ b_{m1} & c_{m2} & \cdots & c_{mk} \end{bmatrix},$$

where
$$c_{ij} = a_{i1}b_{1j} + a_{i2}b_{2j} + \ldots + a_{in}b_{nj} = \sum_{\lambda=1}^{n} a_{i\lambda}b_{\lambda j}$$
$(i = 1, 2, \ldots, m\,; j = 1, 2, \ldots, k)$.

Thus if
$$A = [a_{ij}] = \begin{bmatrix} a_{11} & a_{12} & a_{13} \\ a_{21} & a_{22} & a_{23} \end{bmatrix}, \quad B = [b_i] = \begin{bmatrix} b_1 \\ b_2 \\ b_3 \end{bmatrix},$$

then
$$AB = \begin{bmatrix} a_{11} & a_{12} & a_{13} \\ a_{21} & a_{22} & a_{23} \end{bmatrix} \cdot \begin{bmatrix} b_1 \\ b_2 \\ b_3 \end{bmatrix} = \begin{bmatrix} a_{11}b_1 & a_{12}b_2 & a_{13}b_3 \\ a_{21}b_1 & a_{22}b_2 & a_{23}b_3 \end{bmatrix}.$$

536. Transpose of a Matrix
If the rows and columns of a matrix are interchanged, then the new matrix is called the transpose of the original matrix. If A is the original matrix, its transpose is denoted A^T or $\tilde{A}$.

537. The matrix A is orthogonal if $AA^T = I$.

538. If the matrix product AB is defined, then
$$(AB)^T = B^T A^T.$$

CHAPTER 5. MATRICES AND DETERMINANTS

539. Adjoint of Matrix
If A is a square $n \times n$ matrix, its adjoint, denoted by adj A, is the transpose of the matrix of cofactors C_{ij} of A:
$$\text{adj } A = [C_{ij}]^T.$$

540. Trace of a Matrix
If A is a square $n \times n$ matrix, its trace, denoted by tr A, is defined to be the sum of the terms on the leading diagonal:
tr $A = a_{11} + a_{22} + \ldots + a_{nn}$.

541. Inverse of a Matrix
If A is a square $n \times n$ matrix with a nonsingular determinant det A, then its inverse A^{-1} is given by
$$A^{-1} = \frac{\text{adj } A}{\det A}.$$

542. If the matrix product AB is defined, then
$$(AB)^{-1} = B^{-1}A^{-1}.$$

543. If A is a square $n \times n$ matrix, the eigenvectors X satisfy the equation
$AX = \lambda X$,
while the eigenvalues λ satisfy the characteristic equation
$|A - \lambda I| = 0$.

5.5 Systems of Linear Equations

Variables: $x, y, z, x_1, x_2, \ldots$
Real numbers: $a_1, a_2, a_3, b_1, a_{11}, a_{12}, \ldots$

Determinants: D, D_x, D_y, D_z
Matrices: A, B, X

544. $\begin{cases} a_1 x + b_1 y = d_1 \\ a_2 x + b_2 y = d_2 \end{cases}$,

$x = \dfrac{D_x}{D}$, $y = \dfrac{D_y}{D}$ (Cramer's rule),

where

$D = \begin{vmatrix} a_1 & b_1 \\ a_2 & b_2 \end{vmatrix} = a_1 b_2 - a_2 b_1$,

$D_x = \begin{vmatrix} d_1 & b_1 \\ d_2 & b_2 \end{vmatrix} = d_1 b_2 - d_2 b_1$,

$D_y = \begin{vmatrix} a_1 & d_1 \\ a_2 & d_2 \end{vmatrix} = a_1 d_2 - a_2 d_1$.

545. If $D \neq 0$, then the system has a single solution:

$x = \dfrac{D_x}{D}$, $y = \dfrac{D_y}{D}$.

If $D = 0$ and $D_x \neq 0$ (or $D_y \neq 0$), then the system has no solution.

If $D = D_x = D_y = 0$, then the system has infinitely many solutions.

546. $\begin{cases} a_1 x + b_1 y + c_1 z = d_1 \\ a_2 x + b_2 y + c_2 z = d_2 \\ a_3 x + b_3 y + c_3 z = d_3 \end{cases}$,

$x = \dfrac{D_x}{D}$, $y = \dfrac{D_y}{D}$, $z = \dfrac{D_z}{D}$ (Cramer's rule),

where
$$D = \begin{vmatrix} a_1 & b_1 & c_1 \\ a_2 & b_2 & c_2 \\ a_3 & b_3 & c_3 \end{vmatrix}, \quad D_x = \begin{vmatrix} d_1 & b_1 & c_1 \\ d_2 & b_2 & c_2 \\ d_3 & b_3 & c_3 \end{vmatrix},$$

$$D_y = \begin{vmatrix} a_1 & d_1 & c_1 \\ a_2 & d_2 & c_2 \\ a_3 & d_3 & c_3 \end{vmatrix}, \quad D_z = \begin{vmatrix} a_1 & b_1 & d_1 \\ a_2 & b_2 & d_2 \\ a_3 & b_3 & d_3 \end{vmatrix}.$$

547. If $D \neq 0$, then the system has a single solution:
$$x = \frac{D_x}{D}, \quad y = \frac{D_y}{D}, \quad z = \frac{D_z}{D}.$$
If $D = 0$ and $D_x \neq 0$ (or $D_y \neq 0$ or $D_z \neq 0$), then the system has no solution.
If $D = D_x = D_y = D_z = 0$, then the system has infinitely many solutions.

548. Matrix Form of a System of n Linear Equations in n Unknowns

The set of linear equations
$$\begin{cases} a_{11}x_1 + a_{12}x_2 + \ldots + a_{1n}x_n = b_1 \\ a_{21}x_1 + a_{22}x_2 + \ldots + a_{2n}x_n = b_2 \\ \ldots\ldots\ldots\ldots\ldots\ldots\ldots\ldots\ldots\ldots\ldots \\ a_{n1}x_1 + a_{n2}x_2 + \ldots + a_{nn}x_n = b_n \end{cases}$$
can be written in matrix form
$$\begin{pmatrix} a_{11} & a_{12} & \ldots & a_{1n} \\ a_{21} & a_{22} & \ldots & a_{2n} \\ \vdots & \vdots & & \vdots \\ a_{n1} & a_{n2} & \ldots & a_{nn} \end{pmatrix} \cdot \begin{pmatrix} x_1 \\ x_2 \\ \vdots \\ x_n \end{pmatrix} = \begin{pmatrix} b_1 \\ b_2 \\ \vdots \\ b_n \end{pmatrix},$$
i.e.
$$A \cdot X = B,$$

where

$$A = \begin{pmatrix} a_{11} & a_{12} & \cdots & a_{1n} \\ a_{21} & a_{22} & \cdots & a_{2n} \\ \vdots & \vdots & & \vdots \\ a_{n1} & a_{n2} & \cdots & a_{nn} \end{pmatrix}, \; X = \begin{pmatrix} x_1 \\ x_2 \\ \vdots \\ x_n \end{pmatrix}, \; B = \begin{pmatrix} b_1 \\ b_2 \\ \vdots \\ b_n \end{pmatrix}.$$

549. Solution of a Set of Linear Equations $n \times n$

$$X = A^{-1} \cdot B,$$

where A^{-1} is the inverse of A.

Chapter 6
Vectors

Vectors: $\vec{u}, \vec{v}, \vec{w}, \vec{r}, \vec{AB}, \ldots$
Vector length: $|\vec{u}|, |\vec{v}|, \ldots$
Unit vectors: $\vec{i}, \vec{j}, \vec{k}$
Null vector: $\vec{0}$
Coordinates of vector $\vec{u}$: X_1, Y_1, Z_1
Coordinates of vector $\vec{v}$: X_2, Y_2, Z_2
Scalars: λ, μ
Direction cosines: $\cos\alpha, \cos\beta, \cos\gamma$
Angle between two vectors: θ

6.1 Vector Coordinates

550. Unit Vectors
$\vec{i} = (1, 0, 0),$
$\vec{j} = (0, 1, 0),$
$\vec{k} = (0, 0, 1),$
$|\vec{i}| = |\vec{j}| = |\vec{k}| = 1.$

551. $\vec{r} = \vec{AB} = (x_1 - x_0)\vec{i} + (y_1 - y_0)\vec{j} + (z_1 - z_0)\vec{k}$

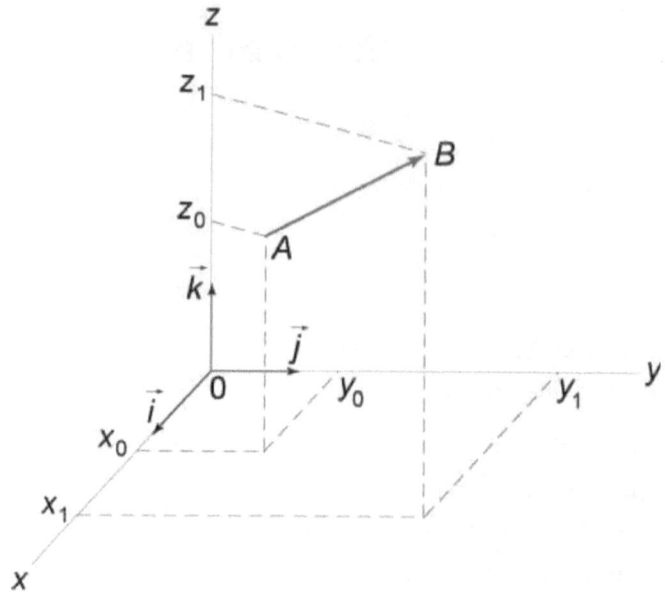

Figure 73.

552. $|\vec{r}| = |\vec{AB}| = \sqrt{(x_1 - x_0)^2 + (y_1 - y_0)^2 + (z_1 - z_0)^2}$

553. If $\vec{AB} = \vec{r}$, then $\vec{BA} = -\vec{r}$.

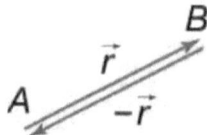

Figure 74.

554. $X = |\vec{r}| \cos\alpha$,
$Y = |\vec{r}| \cos\beta$,
$Z = |\vec{r}| \cos\gamma$.

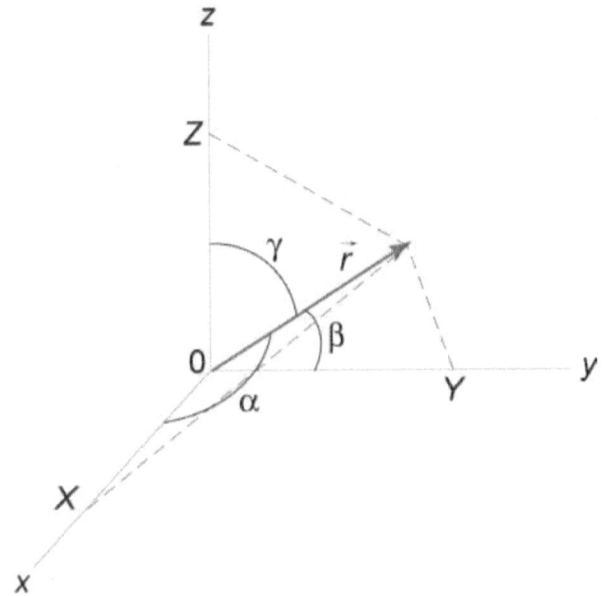

Figure 75.

555. If $\vec{r}(X, Y, Z) = \vec{r}_1(X_1, Y_1, Z_1)$, then $X = X_1$, $Y = Y_1$, $Z = Z_1$.

6.2 Vector Addition

556. $\vec{w} = \vec{u} + \vec{v}$

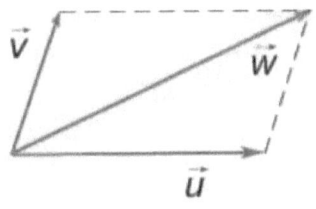

Figure 76.

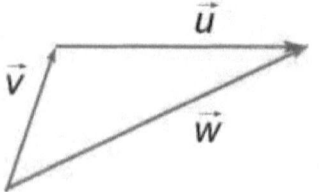

Figure 77.

557. $\vec{w} = \vec{u}_1 + \vec{u}_2 + \vec{u}_3 + \ldots + \vec{u}_n$

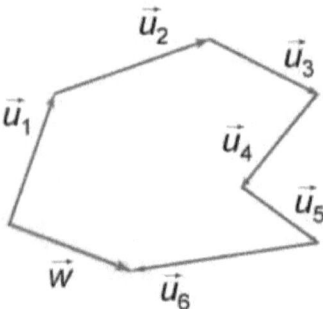

Figure 78.

558. Commutative Law
$\vec{u} + \vec{v} = \vec{v} + \vec{u}$

559. Associative Law
$(\vec{u} + \vec{v}) + \vec{w} = \vec{u} + (\vec{v} + \vec{w})$

560. $\vec{u} + \vec{v} = (X_1 + X_2, Y_1 + Y_2, Z_1 + Z_2)$

6.3 Vector Subtraction

561. $\vec{w} = \vec{u} - \vec{v}$ if $\vec{v} + \vec{w} = \vec{u}$.

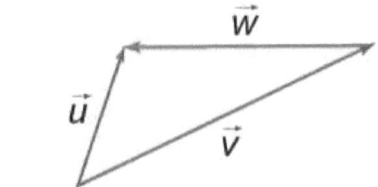

Figure 79.

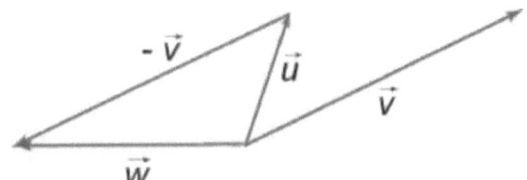

Figure 80.

562. $\vec{u} - \vec{v} = \vec{u} + (-\vec{v})$

563. $\vec{u} - \vec{u} = \vec{0} = (0, 0, 0)$

564. $|\vec{0}| = 0$

565. $\vec{u} - \vec{v} = (X_1 - X_2, Y_1 - Y_2, Z_1 - Z_2)$,

6.4 Scaling Vectors

566. $\vec{w} = \lambda \vec{u}$

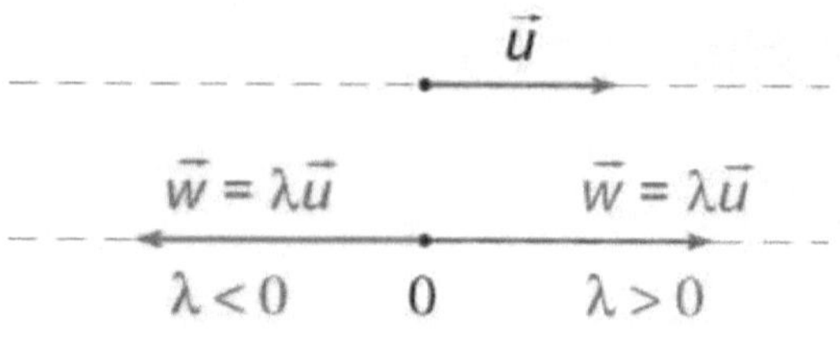

Figure 81.

567. $|\vec{w}| = |\lambda| \cdot |\vec{u}|$

568. $\lambda \vec{u} = (\lambda X, \lambda Y, \lambda Z)$

569. $\lambda \vec{u} = \vec{u} \lambda$

570. $(\lambda + \mu)\vec{u} = \lambda \vec{u} + \mu \vec{u}$

571. $\lambda(\mu \vec{u}) = \mu(\lambda \vec{u}) = (\lambda \mu)\vec{u}$

572. $\lambda(\vec{u} + \vec{v}) = \lambda \vec{u} + \lambda \vec{v}$

6.5 Scalar Product

573. Scalar Product of Vectors $\vec{u}$ and $\vec{v}$
$$\vec{u} \cdot \vec{v} = |\vec{u}| \cdot |\vec{v}| \cdot \cos \theta,$$
where θ is the angle between vectors $\vec{u}$ and $\vec{v}$.

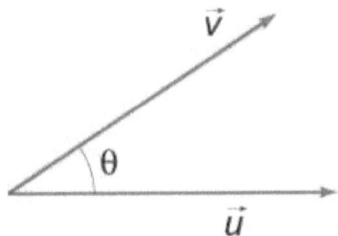

Figure 82.

574. Scalar Product in Coordinate Form
If $\vec{u} = (X_1, Y_1, Z_1)$, $\vec{v} = (X_2, Y_2, Z_2)$, then
$\vec{u} \cdot \vec{v} = X_1 X_2 + Y_1 Y_2 + Z_1 Z_2$.

575. Angle Between Two Vectors
If $\vec{u} = (X_1, Y_1, Z_1)$, $\vec{v} = (X_2, Y_2, Z_2)$, then
$$\cos\theta = \frac{X_1 X_2 + Y_1 Y_2 + Z_1 Z_2}{\sqrt{X_1^2 + Y_1^2 + Z_1^2}\sqrt{X_2^2 + Y_2^2 + Z_2^2}}.$$

576. Commutative Property
$\vec{u} \cdot \vec{v} = \vec{v} \cdot \vec{u}$

577. Associative Property
$(\lambda\vec{u}) \cdot (\mu\vec{v}) = \lambda\mu\vec{u} \cdot \vec{v}$

578. Distributive Property
$\vec{u} \cdot (\vec{v} + \vec{w}) = \vec{u} \cdot \vec{v} + \vec{u} \cdot \vec{w}$

579. $\vec{u} \cdot \vec{v} = 0$ if $\vec{u}, \vec{v}$ are orthogonal ($\theta = \frac{\pi}{2}$).

580. $\vec{u} \cdot \vec{v} > 0$ if $0 < \theta < \frac{\pi}{2}$.

581. $\vec{u} \cdot \vec{v} < 0$ if $\dfrac{\pi}{2} < \theta < \pi$.

582. $\vec{u} \cdot \vec{v} \le |\vec{u}| \cdot |\vec{v}|$

583. $\vec{u} \cdot \vec{v} = |\vec{u}| \cdot |\vec{v}|$ if $\vec{u}, \vec{v}$ are parallel ($\theta = 0$).

584. If $\vec{u} = (X_1, Y_1, Z_1)$, then
$$\vec{u} \cdot \vec{u} = \vec{u}^2 = |\vec{u}|^2 = X_1^2 + Y_1^2 + Z_1^2.$$

585. $\vec{i} \cdot \vec{i} = \vec{j} \cdot \vec{j} = \vec{k} \cdot \vec{k} = 1$

586. $\vec{i} \cdot \vec{j} = \vec{j} \cdot \vec{k} = \vec{k} \cdot \vec{i} = 0$

6.6 Vector Product

587. Vector Product of Vectors $\vec{u}$ and $\vec{v}$
$\vec{u} \times \vec{v} = \vec{w}$, where
- $|\vec{w}| = |\vec{u}| \cdot |\vec{v}| \cdot \sin\theta$, where $0 \le \theta \le \dfrac{\pi}{2}$;
  and $\vec{w} \perp \vec{v}$;
- Vectors $\vec{u}, \vec{v}, \vec{w}$ form a right-handed screw.

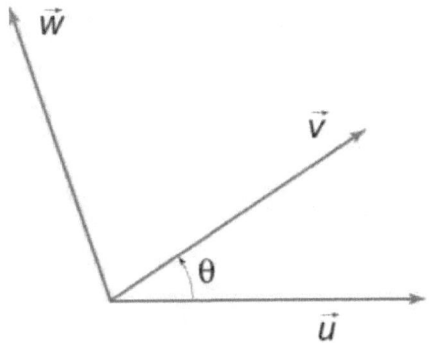

Figure 83.

588. $\vec{w} = \vec{u} \times \vec{v} = \begin{vmatrix} \vec{i} & \vec{j} & \vec{k} \\ X_1 & Y_1 & Z_1 \\ X_2 & Y_2 & Z_2 \end{vmatrix}$

589. $\vec{w} = \vec{u} \times \vec{v} = \left(\begin{vmatrix} Y_1 & Z_1 \\ Y_2 & Z_2 \end{vmatrix}, -\begin{vmatrix} X_1 & Z_1 \\ X_2 & Z_2 \end{vmatrix}, \begin{vmatrix} X_1 & Y_1 \\ X_2 & Y_2 \end{vmatrix} \right)$

590. $S = |\vec{u} \times \vec{v}| = |\vec{u}| \cdot |\vec{v}| \cdot \sin \theta$ (Fig.83)

591. Angle Between Two Vectors (Fig.83)
$\sin \theta = \dfrac{\vec{u} \times \vec{v}}{|\vec{u}| \cdot |\vec{v}|}$

592. Noncommutative Property
$\vec{u} \times \vec{v} = -(\vec{v} \times \vec{u})$

593. Associative Property
$(\lambda \vec{u}) \times (\mu \vec{v}) = \lambda \mu \vec{u} \times \vec{v}$

594. Distributive Property
$$\vec{u} \times (\vec{v} + \vec{w}) = \vec{u} \times \vec{v} + \vec{u} \times \vec{w}$$

595. $\vec{u} \times \vec{v} = \vec{0}$ if $\vec{u}$ and $\vec{v}$ are parallel ($\theta = 0$).

596. $\vec{i} \times \vec{i} = \vec{j} \times \vec{j} = \vec{k} \times \vec{k} = \vec{0}$

597. $\vec{i} \times \vec{j} = \vec{k}, \ \vec{j} \times \vec{k} = \vec{i}, \ \vec{k} \times \vec{i} = \vec{j}$

6.7 Triple Product

598. Scalar Triple Product
$$[\vec{u}\vec{v}\vec{w}] = \vec{u} \cdot (\vec{v} \times \vec{w}) = \vec{v} \cdot (\vec{w} \times \vec{u}) = \vec{w} \cdot (\vec{u} \times \vec{v})$$

599. $[\vec{u}\vec{v}\vec{w}] = [\vec{w}\vec{u}\vec{v}] = [\vec{v}\vec{w}\vec{u}] = -[\vec{v}\vec{u}\vec{w}] = -[\vec{w}\vec{v}\vec{u}] = -[\vec{u}\vec{w}\vec{v}]$

600. $k\vec{u} \cdot (\vec{v} \times \vec{w}) = k[\vec{u}\vec{v}\vec{w}]$

601. Scalar Triple Product in Coordinate Form
$$\vec{u} \cdot (\vec{v} \times \vec{w}) = \begin{vmatrix} X_1 & Y_1 & Z_1 \\ X_2 & Y_2 & Z_2 \\ X_3 & Y_3 & Z_3 \end{vmatrix},$$
where
$\vec{u} = (X_1, Y_1, Z_1), \ \vec{v} = (X_2, Y_2, Z_2), \ \vec{w} = (X_3, Y_3, Z_3)$.

602. Volume of Parallelepiped
$$V = |\vec{u} \cdot (\vec{v} \times \vec{w})|$$

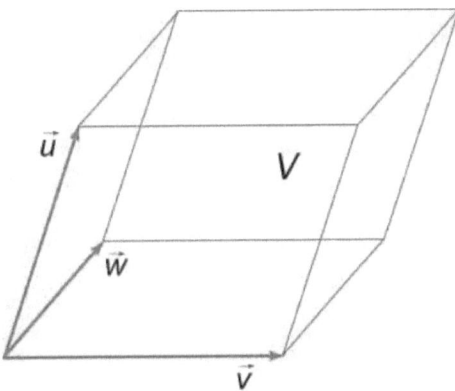

Figure 84.

603. Volume of Pyramid
$$V = \frac{1}{6}\left|\vec{u}\cdot(\vec{v}\times\vec{w})\right|$$

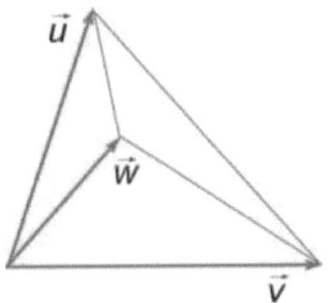

Figure 85.

604. If $\vec{u}\cdot(\vec{v}\times\vec{w})=0$, then the vectors $\vec{u}$, $\vec{v}$, and $\vec{w}$ are linearly dependent, so $\vec{w}=\lambda\vec{u}+\mu\vec{v}$ for some scalars λ and μ.

605. If $\vec{u}\cdot(\vec{v}\times\vec{w})\neq 0$, then the vectors $\vec{u}$, $\vec{v}$, and $\vec{w}$ are linearly independent.

606. Vector Triple Product
$$\vec{u} \times (\vec{v} \times \vec{w}) = (\vec{u} \cdot \vec{w})\vec{v} - (\vec{u} \cdot \vec{v})\vec{w}$$

Chapter 7
Analytic Geometry

7.1 One-Dimensional Coordinate System

Point coordinates: $x_0, x_1, x_2, y_0, y_1, y_2$
Real number: λ
Distance between two points: d

607. Distance Between Two Points
$$d = AB = |x_2 - x_1| = |x_1 - x_2|$$

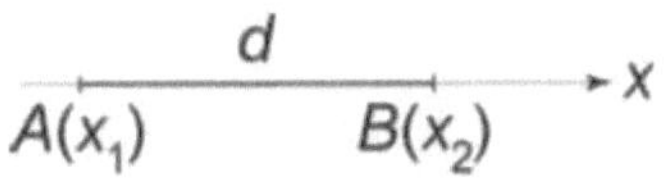

Figure 86.

608. Dividing a Line Segment in the Ratio λ
$$x_0 = \frac{x_1 + \lambda x_2}{1 + \lambda}, \quad \lambda = \frac{AC}{CB}, \quad \lambda \neq -1.$$

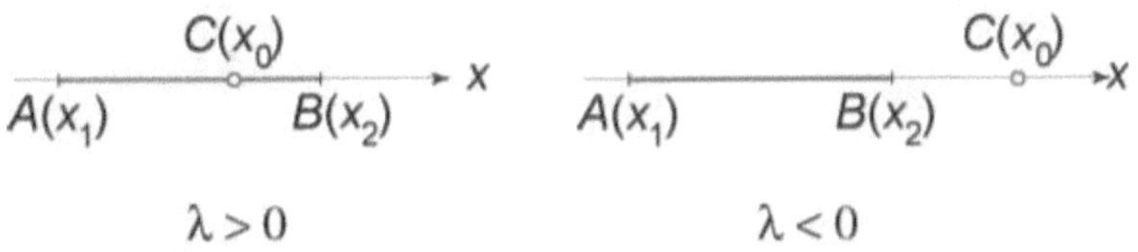

$\lambda > 0$ $\lambda < 0$

Figure 87.

609. Midpoint of a Line Segment

$$x_0 = \frac{x_1 + x_2}{2}, \quad \lambda = 1.$$

7.2 Two-Dimensional Coordinate System

Point coordinates: x_0, x_1, x_2, y_0, y_1, y_2
Polar coordinates: r, φ
Real number: λ
Positive real numbers: a, b, c,
Distance between two points: d
Area: S

610. Distance Between Two Points

$$d = AB = \sqrt{(x_2 - x_1)^2 + (y_2 - y_1)^2}$$

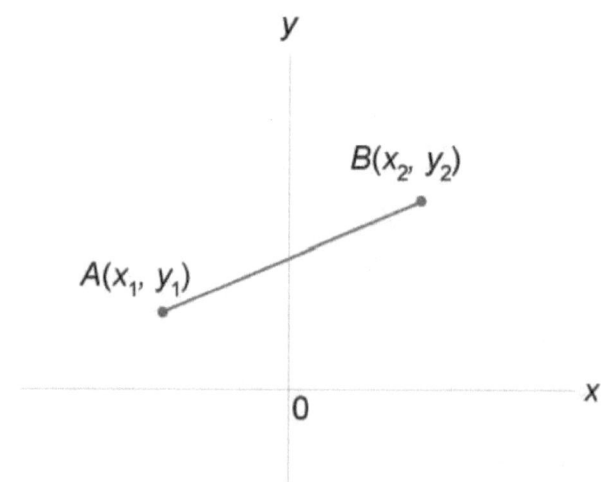

Figure 88.

611. Dividing a Line Segment in the Ratio λ

$$x_0 = \frac{x_1 + \lambda x_2}{1+\lambda}, \quad y_0 = \frac{y_1 + \lambda y_2}{1+\lambda},$$

$$\lambda = \frac{AC}{CB}, \quad \lambda \neq -1.$$

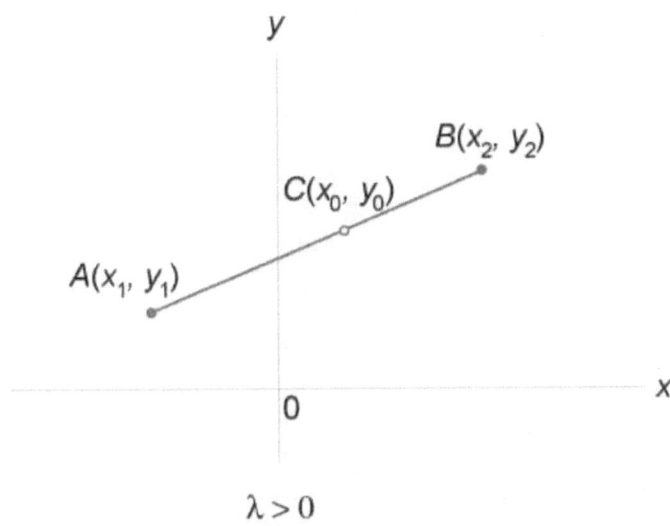

$\lambda > 0$

Figure 89.

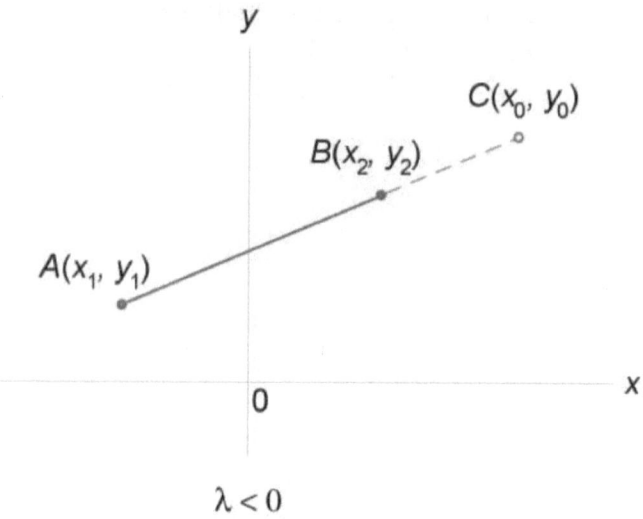

$\lambda < 0$

Figure 90.

612. Midpoint of a Line Segment

$$x_0 = \frac{x_1 + x_2}{2}, \; y_0 = \frac{y_1 + y_2}{2}, \; \lambda = 1.$$

613. Centroid (Intersection of Medians) of a Triangle

$$x_0 = \frac{x_1 + x_2 + x_3}{3}, \; y_0 = \frac{y_1 + y_2 + y_3}{3},$$

where $A(x_1,y_1)$, $B(x_2,y_2)$, and $C(x_3,y_3)$ are vertices of the triangle ABC.

CHAPTER 7. ANALYTIC GEOMETRY

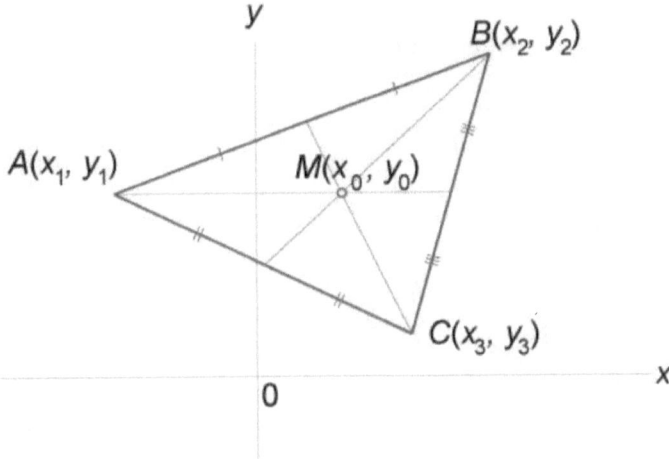

Figure 91.

614. Incenter (Intersection of Angle Bisectors) of a Triangle

$$x_0 = \frac{ax_1 + bx_2 + cx_3}{a+b+c}, \quad y_0 = \frac{ay_1 + by_2 + cy_3}{a+b+c},$$

where $a = BC$, $b = CA$, $c = AB$.

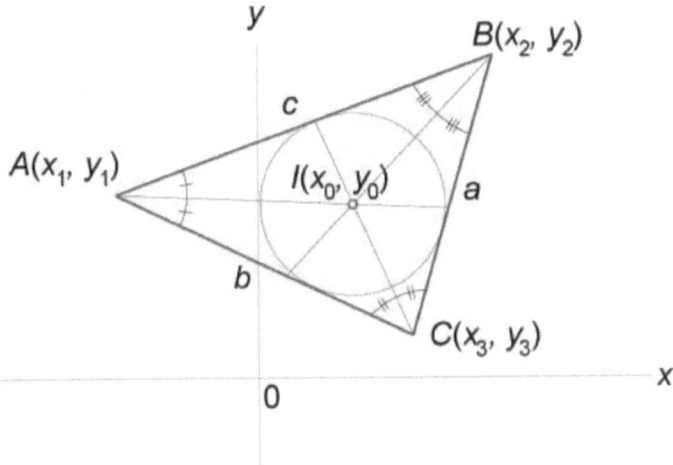

Figure 92.

CHAPTER 7. ANALYTIC GEOMETRY

615. Circumcenter (Intersection of the Side Perpendicular Bisectors) of a Triangle

$$x_0 = \frac{\begin{vmatrix} x_1^2+y_1^2 & y_1 & 1 \\ x_2^2+y_2^2 & y_2 & 1 \\ x_3^2+y_3^2 & y_3 & 1 \end{vmatrix}}{2\begin{vmatrix} x_1 & y_1 & 1 \\ x_2 & y_2 & 1 \\ x_3 & y_3 & 1 \end{vmatrix}}, \quad y_0 = \frac{\begin{vmatrix} x_1 & x_1^2+y_1^2 & 1 \\ x_2 & x_2^2+y_2^2 & 1 \\ x_3 & x_3^2+y_3^2 & 1 \end{vmatrix}}{2\begin{vmatrix} x_1 & y_1 & 1 \\ x_2 & y_2 & 1 \\ x_3 & y_3 & 1 \end{vmatrix}}$$

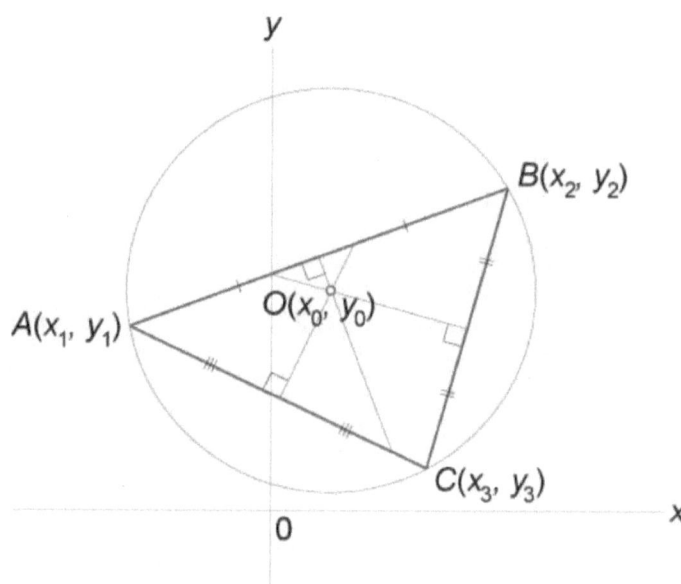

Figure 93.

616. Orthocenter (Intersection of Altitudes) of a Triangle

$$x_0 = \frac{\begin{vmatrix} y_1 & x_2x_3+y_1^2 & 1 \\ y_2 & x_3x_1+y_2^2 & 1 \\ y_3 & x_1x_2+y_3^2 & 1 \end{vmatrix}}{\begin{vmatrix} x_1 & y_1 & 1 \\ x_2 & y_2 & 1 \\ x_3 & y_3 & 1 \end{vmatrix}}, \quad y_0 = \frac{\begin{vmatrix} x_1^2+y_2y_3 & x_1 & 1 \\ x_2^2+y_3y_1 & x_2 & 1 \\ x_3^2+y_1y_2 & x_3 & 1 \end{vmatrix}}{\begin{vmatrix} x_1 & y_1 & 1 \\ x_2 & y_2 & 1 \\ x_3 & y_3 & 1 \end{vmatrix}}$$

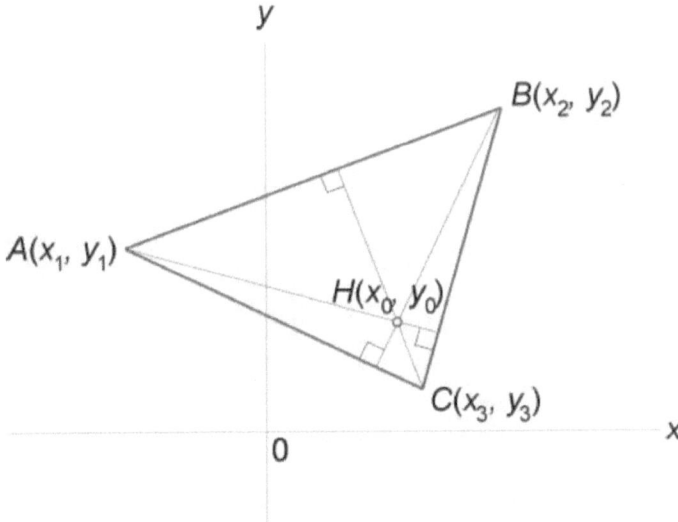

Figure 94.

617. Area of a Triangle

$$S = (\pm)\frac{1}{2}\begin{vmatrix} x_1 & y_1 & 1 \\ x_2 & y_2 & 1 \\ x_3 & y_3 & 1 \end{vmatrix} = (\pm)\frac{1}{2}\begin{vmatrix} x_2-x_1 & y_2-y_1 \\ x_3-x_1 & y_3-y_1 \end{vmatrix}$$

618. Area of a Quadrilateral

$$S = (\pm)\frac{1}{2}[(x_1 - x_2)(y_1 + y_2) + (x_2 - x_3)(y_2 + y_3) + (x_3 - x_4)(y_3 + y_4) + (x_4 - x_1)(y_4 + y_1)]$$

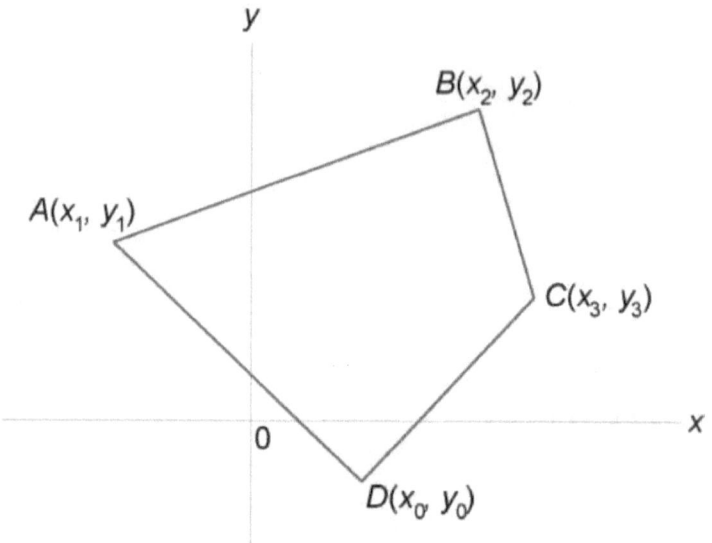

Figure 95.

Note: In formulas 617, 618 we choose the sign (+) or (−) so that to get a positive answer for area.

619. Distance Between Two Points in Polar Coordinates

$$d = AB = \sqrt{r_1^2 + r_2^2 - 2r_1 r_2 \cos(\varphi_2 - \varphi_1)}$$

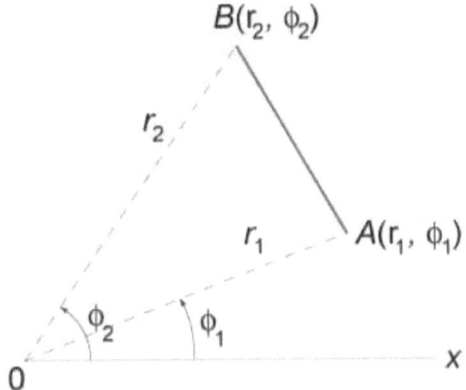

Figure 96.

620. Converting Rectangular Coordinates to Polar Coordinates

$x = r \cos \varphi$, $y = r \sin \varphi$.

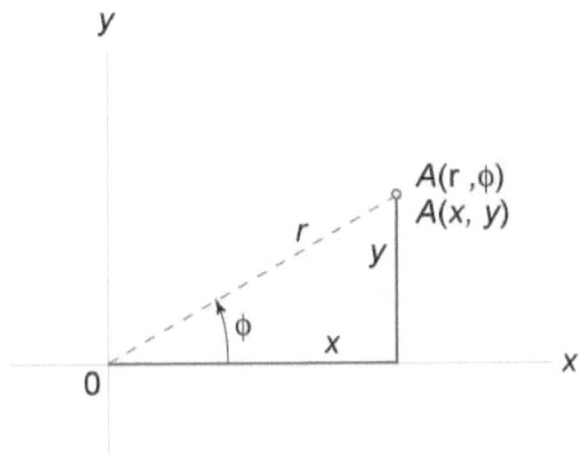

Figure 97.

621. Converting Polar Coordinates to Rectangular Coordinates

$r = \sqrt{x^2 + y^2}$, $\tan \varphi = \dfrac{y}{x}$.

7.3 Straight Line in Plane

Point coordinates: X, Y, x, x_0, x_1, y_0, y_1, a_1, a_2, ...
Real numbers: k, a, b, p, t, A, B, C, A_1, A_2, ...
Angles: α, β
Angle between two lines: φ
Normal vector: $\vec{n}$
Position vectors: $\vec{r}$, $\vec{a}$, $\vec{b}$

622. General Equation of a Straight Line
$Ax + By + C = 0$

623. Normal Vector to a Straight Line
The vector $\vec{n}(A, B)$ is normal to the line $Ax + By + C = 0$.

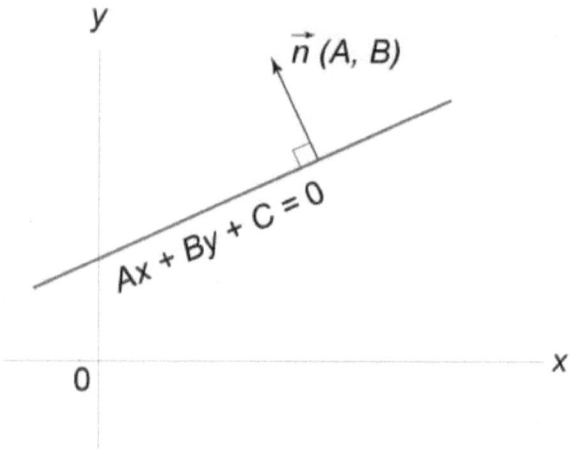

Figure 98.

624. Explicit Equation of a Straight Line (Slope-Intercept Form)
$y = kx + b$.

The gradient of the line is $k = \tan \alpha$.

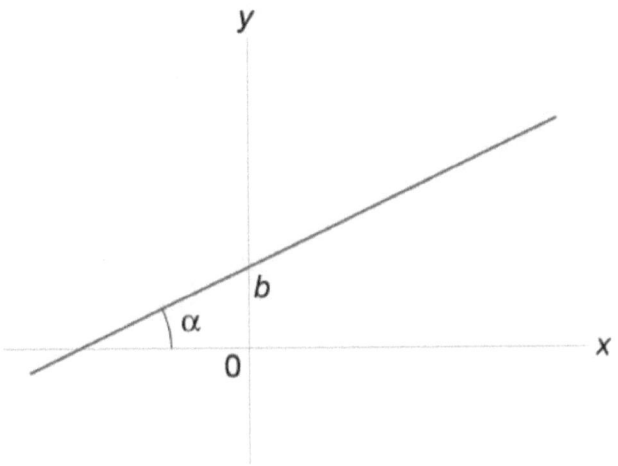

Figure 99.

625. Gradient of a Line

$$k = \tan \alpha = \frac{y_2 - y_1}{x_2 - x_1}$$

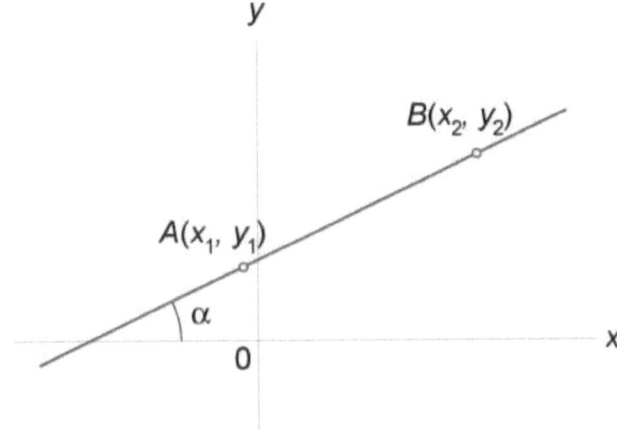

Figure 100.

CHAPTER 7. ANALYTIC GEOMETRY

626. Equation of a Line Given a Point and the Gradient
$$y = y_0 + k(x - x_0),$$
where k is the gradient, $P(x_0, y_0)$ is a point on the line.

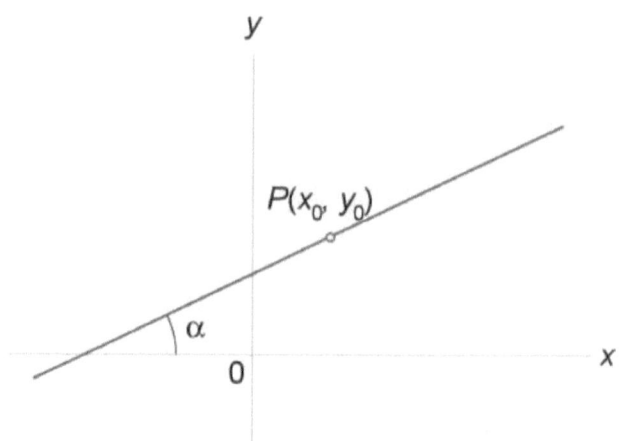

Figure 101.

627. Equation of a Line That Passes Through Two Points
$$\frac{y - y_1}{y_2 - y_1} = \frac{x - x_1}{x_2 - x_1}$$
or
$$\begin{vmatrix} x & y & 1 \\ x_1 & y_1 & 1 \\ x_2 & y_2 & 1 \end{vmatrix} = 0.$$

CHAPTER 7. ANALYTIC GEOMETRY

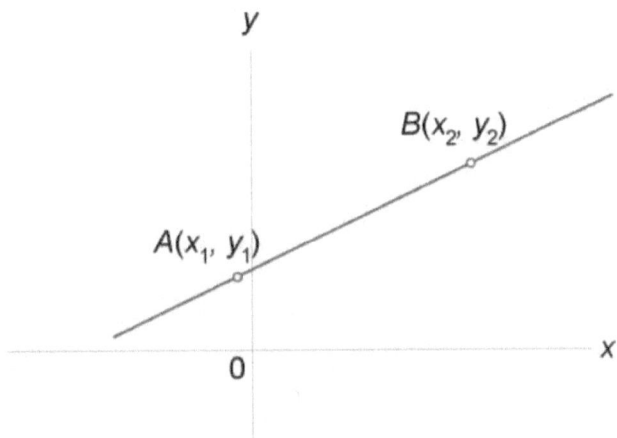

Figure 102.

628. Intercept Form

$$\frac{x}{a}+\frac{y}{b}=1$$

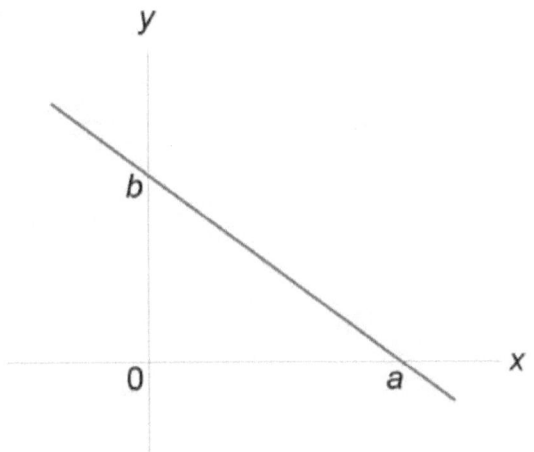

Figure 103.

CHAPTER 7. ANALYTIC GEOMETRY

629. Normal Form

$$x \cos \beta + y \sin \beta - p = 0$$

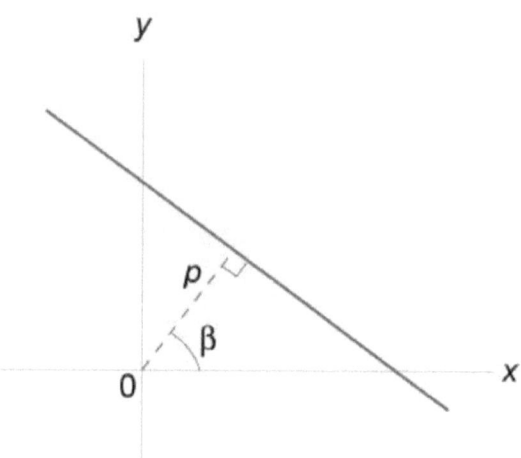

Figure 104.

630. Point Direction Form

$$\frac{x - x_1}{X} = \frac{y - y_1}{Y},$$

where (X, Y) is the direction of the line and $P_1(x_1, y_1)$ lies on the line.

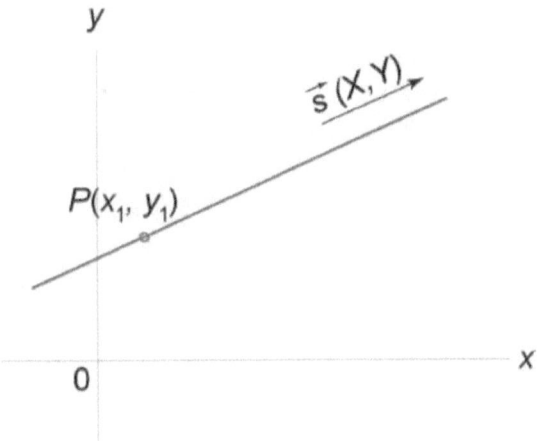

Figure 105.

631. Vertical Line
$x = a$

632. Horizontal Line
$y = b$

633. Vector Equation of a Straight Line
$\vec{r} = \vec{a} + t\vec{b}$,
where
O is the origin of the coordinates,
X is any variable point on the line,
$\vec{a}$ is the position vector of a known point A on the line,
$\vec{b}$ is a known vector of direction, parallel to the line,
t is a parameter,

$\vec{r} = \overrightarrow{OX}$ is the position vector of any point X on the line.

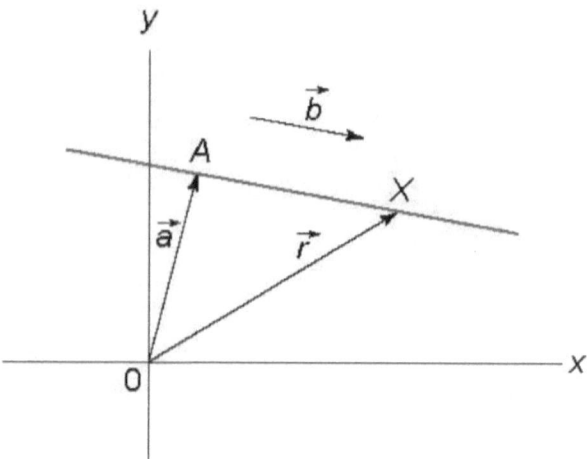

Figure 106.

634. Straight Line in Parametric Form
$$\begin{cases} x = a_1 + tb_1 \\ y = a_2 + tb_2 \end{cases},$$
where
(x, y) are the coordinates of any unknown point on the line,
(a_1, a_2) are the coordinates of a known point on the line,
(b_1, b_2) are the coordinates of a vector parallel to the line,
t is a parameter.

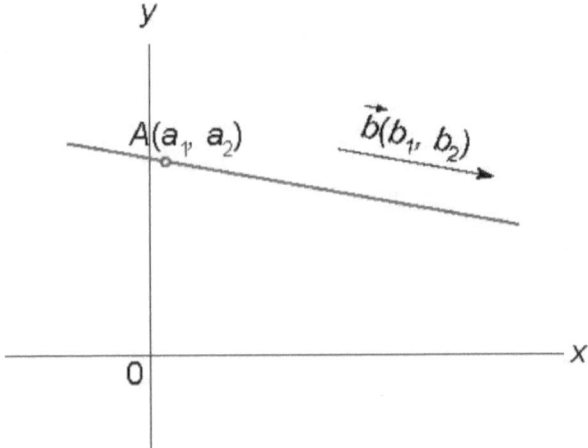

Figure 107.

635. Distance From a Point To a Line

The distance from the point $P(a, b)$ to the line $Ax + By + C = 0$ is

$$d = \frac{|Aa + Bb + C|}{\sqrt{A^2 + B^2}}.$$

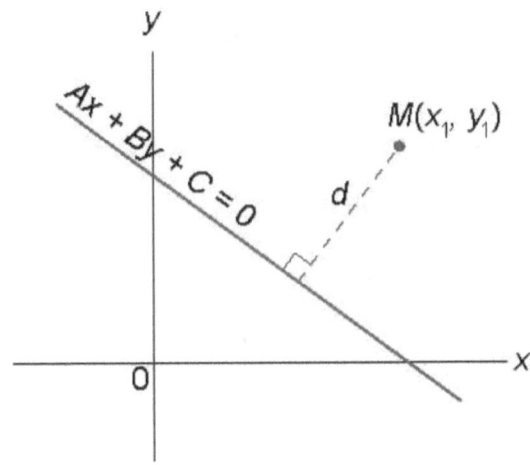

Figure 108.

636. Parallel Lines

Two lines $y = k_1 x + b_1$ and $y = k_2 x + b_2$ are parallel if $k_1 = k_2$.

Two lines $A_1 x + B_1 y + C_1 = 0$ and $A_2 x + B_2 y + C_2 = 0$ are parallel if

$$\frac{A_1}{A_2} = \frac{B_1}{B_2}.$$

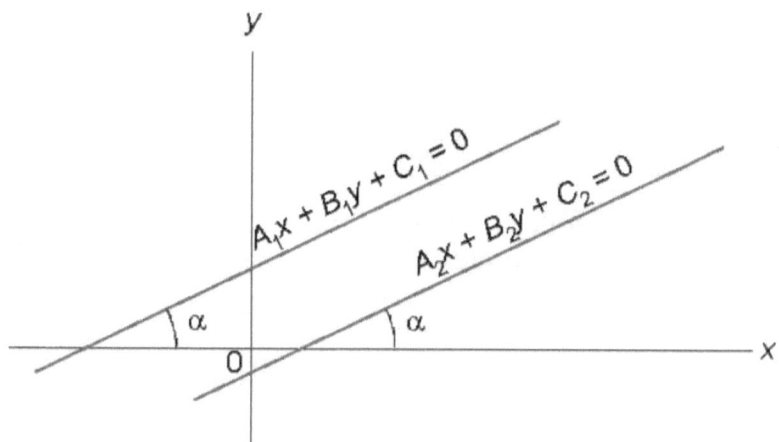

Figure 109.

637. Perpendicular Lines

Two lines $y = k_1 x + b_1$ and $y = k_2 x + b_2$ are perpendicular if $k_2 = -\frac{1}{k_1}$ or, equivalently, $k_1 k_2 = -1$.

Two lines $A_1 x + B_1 y + C_1 = 0$ and $A_2 x + B_2 y + C_2 = 0$ are perpendicular if
$A_1 A_2 + B_1 B_2 = 0$.

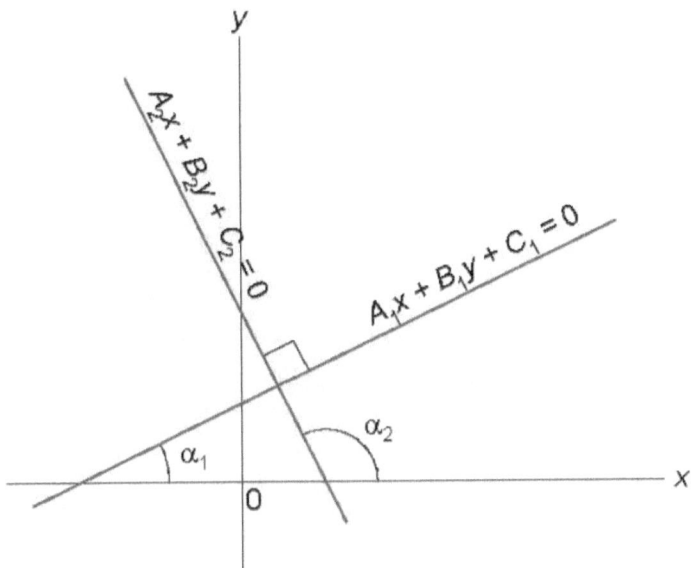

Figure 110.

638. Angle Between Two Lines

$$\tan\varphi = \frac{k_2 - k_1}{1 + k_1 k_2},$$

$$\cos\varphi = \frac{A_1 A_2 + B_1 B_2}{\sqrt{A_1^2 + B_1^2} \cdot \sqrt{A_2^2 + B_2^2}}.$$

CHAPTER 7. ANALYTIC GEOMETRY

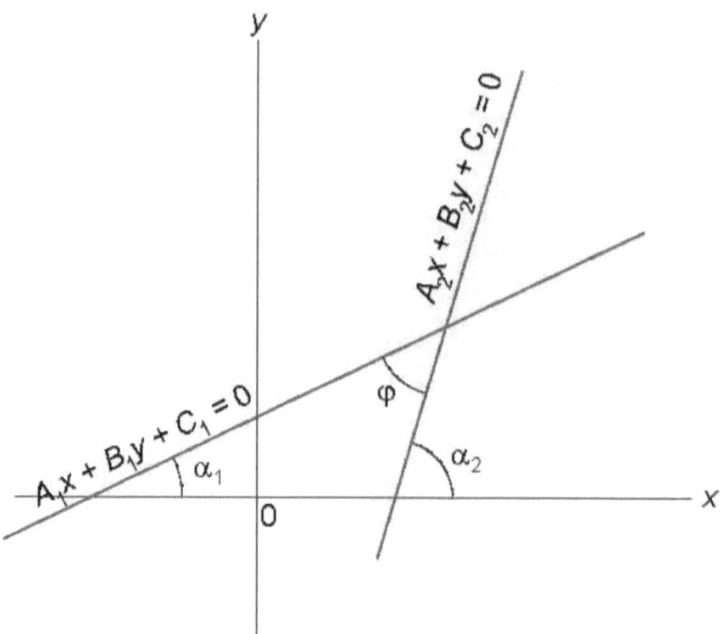

Figure 111.

639. Intersection of Two Lines

If two lines $A_1x + B_1y + C_1 = 0$ and $A_2x + B_2y + C_2 = 0$ intersect, the intersection point has coordinates

$$x_0 = \frac{-C_1B_2 + C_2B_1}{A_1B_2 - A_2B_1}, \quad y_0 = \frac{-A_1C_2 + A_2C_1}{A_1B_2 - A_2B_1}.$$

7.4 Circle

Radius: R
Center of circle: (a, b)
Point coordinates: $x, y, x_1, y_1, \ldots$
Real numbers: A, B, C, D, E, F, t

640. Equation of a Circle Centered at the Origin (Standard Form)

$$x^2 + y^2 = R^2$$

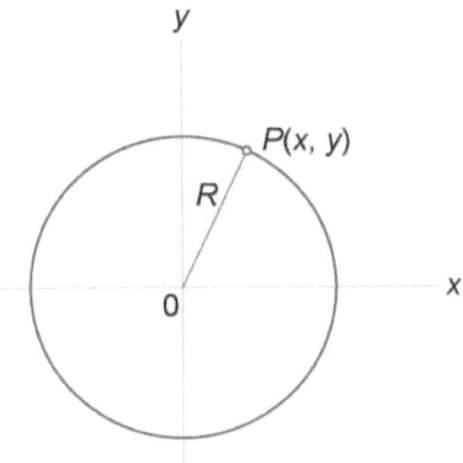

Figure 112.

641. Equation of a Circle Centered at Any Point (a, b)

$$(x-a)^2 + (y-b)^2 = R^2$$

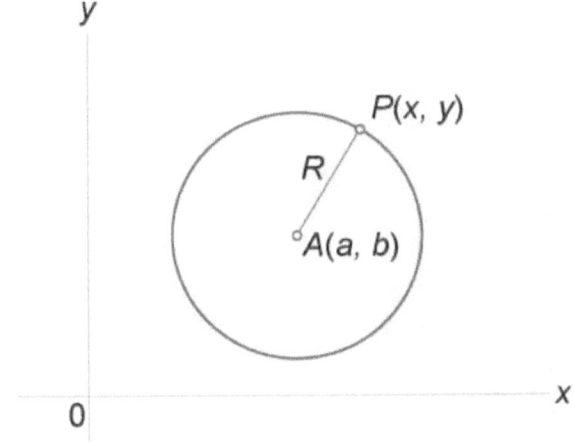

Figure 113.

642. Three Point Form

$$\begin{vmatrix} x^2+y^2 & x & y & 1 \\ x_1^2+y_1^2 & x_1 & y_1 & 1 \\ x_2^2+y_2^2 & x_2 & y_2 & 1 \\ x_3^2+y_3^2 & x_3 & y_3 & 1 \end{vmatrix} = 0$$

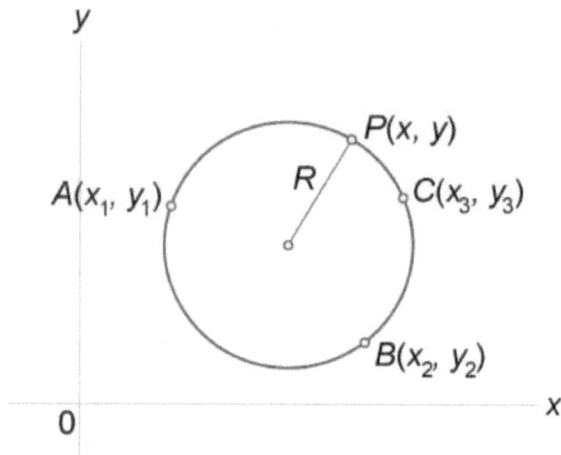

Figure 114.

643. Parametric Form

$$\begin{cases} x = R\cos t \\ y = R\sin t \end{cases}, \ 0 \le t \le 2\pi.$$

644. General Form

$Ax^2 + Ay^2 + Dx + Ey + F = 0$ (A nonzero, $D^2 + E^2 > 4AF$).
The center of the circle has coordinates (a, b), where

$$a = -\frac{D}{2A}, \ b = -\frac{E}{2A}.$$

The radius of the circle is

$$R = \sqrt{\frac{D^2 + E^2 - 4AF}{2|A|}}.$$

7.5 Ellipse

Semimajor axis: a
Semiminor axis: b
Foci: $F_1(-c, 0)$, $F_2(c, 0)$
Distance between the foci: 2c
Eccentricity: e
Real numbers: A, B, C, D, E, F, t
Perimeter: L
Area: S

645. Equation of an Ellipse (Standard Form)
$$\frac{x^2}{a^2} + \frac{y^2}{b^2} = 1$$

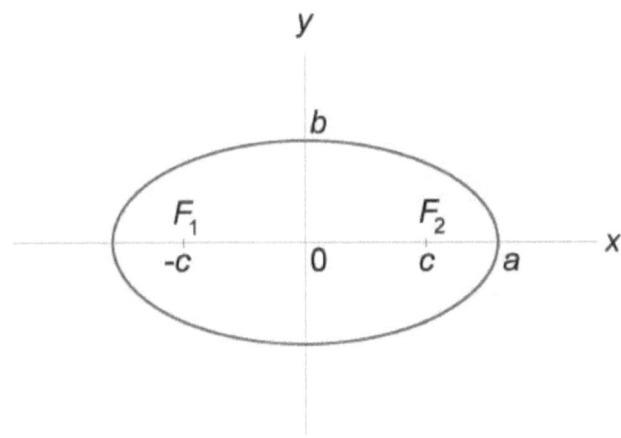

Figure 115.

646. $r_1 + r_2 = 2a$,

where r_1, r_2 are distances from any point $P(x,y)$ on the ellipse to the two foci.

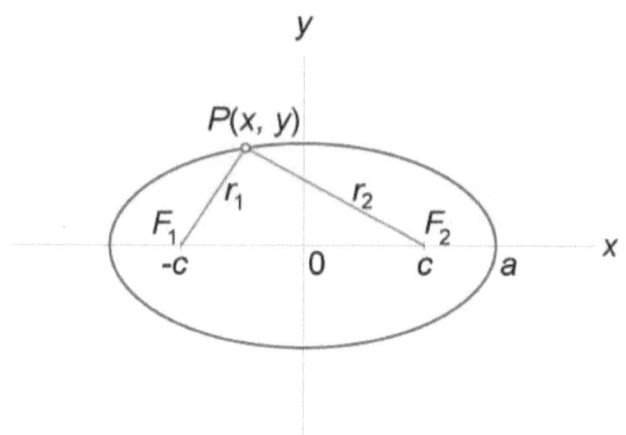

Figure 116.

647. $a^2 = b^2 + c^2$

648. Eccentricity

$$e = \frac{c}{a} < 1$$

649. Equations of Directrices

$$x = \pm \frac{a}{e} = \pm \frac{a^2}{c}$$

650. Parametric Form

$$\begin{cases} x = a\cos t \\ y = b\sin t \end{cases}, \quad 0 \le t \le 2\pi.$$

651. **General Form**
$$Ax^2 + Bxy + Cy^2 + Dx + Ey + F = 0,$$
where $B^2 - 4AC < 0$.

652. **General Form with Axes Parallel to the Coordinate Axes**
$$Ax^2 + Cy^2 + Dx + Ey + F = 0,$$
where $AC > 0$.

653. **Circumference**
$$L = 4aE(e),$$
where the function E is the complete elliptic integral of the second kind.

654. **Approximate Formulas of the Circumference**
$$L = \pi \left(1.5(a+b) - \sqrt{ab}\right),$$
$$L = \pi \sqrt{2(a^2 + b^2)}.$$

655. $S = \pi ab$

7.6 Hyperbola

Transverse axis: a
Conjugate axis: b
Foci: $F_1(-c, 0)$, $F_2(c, 0)$
Distance between the foci: 2c
Eccentricity: e
Asymptotes: s, t
Real numbers: A, B, C, D, E, F, t, k

CHAPTER 7. ANALYTIC GEOMETRY

656. Equation of a Hyperbola (Standard Form)
$$\frac{x^2}{a^2} - \frac{y^2}{b^2} = 1$$

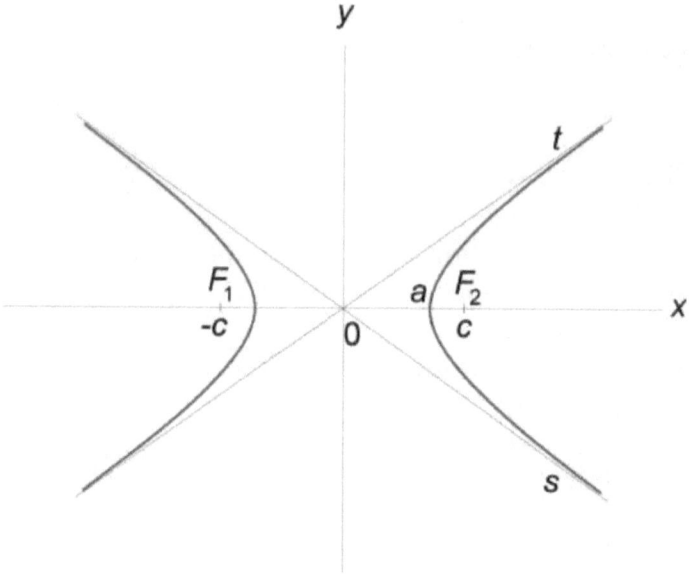

Figure 117.

657. $|r_1 - r_2| = 2a$,

where r_1, r_2 are distances from any point $P(x,y)$ on the hyperbola to the two foci.

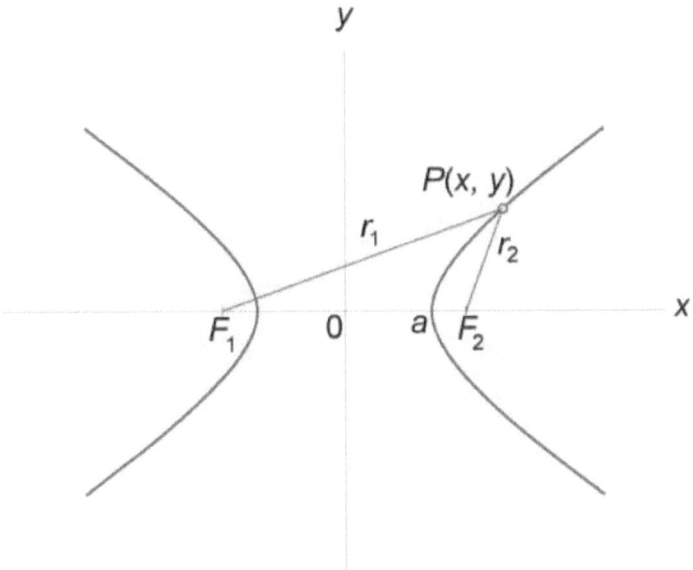

Figure 118.

658. Equations of Asymptotes

$$y = \pm \frac{b}{a} x$$

659. $c^2 = a^2 + b^2$

660. Eccentricity

$$e = \frac{c}{a} > 1$$

661. Equations of Directrices

$$x = \pm \frac{a}{e} = \pm \frac{a^2}{c}$$

662. Parametric Equations of the Right Branch of a Hyperbola
$$\begin{cases} x = a\cosh t \\ y = b\sinh t \end{cases}, \quad 0 \le t \le 2\pi.$$

663. General Form
$$Ax^2 + Bxy + Cy^2 + Dx + Ey + F = 0,$$
where $B^2 - 4AC > 0$.

664. General Form with Axes Parallel to the Coordinate Axes
$$Ax^2 + Cy^2 + Dx + Ey + F = 0,$$
where $AC < 0$.

665. Asymptotic Form
$$xy = \frac{e^2}{4},$$
or
$$y = \frac{k}{x}, \text{ where } k = \frac{e^2}{4}.$$
In this case, the asymptotes have equations $x = 0$ and $y = 0$.

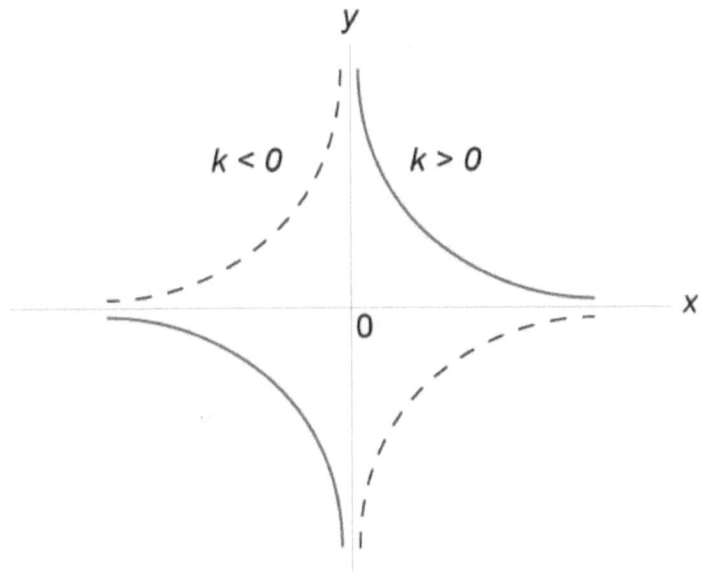

Figure 119.

7.7 Parabola

Focal parameter: p
Focus: F
Vertex: $M(x_0, y_0)$
Real numbers: A, B, C, D, E, F, p, a, b, c

666. Equation of a Parabola (Standard Form)
$$y^2 = 2px$$

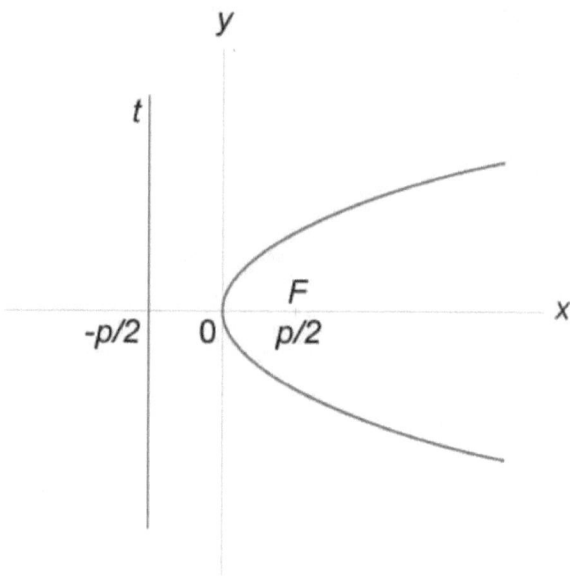

Figure 120.

Equation of the directrix
$$x = -\frac{p}{2},$$
Coordinates of the focus
$$F\left(\frac{p}{2}, 0\right),$$
Coordinates of the vertex
$M(0, 0)$.

667. General Form
$$Ax^2 + Bxy + Cy^2 + Dx + Ey + F = 0,$$
where $B^2 - 4AC = 0$.

668. $y = ax^2$, $p = \dfrac{1}{2a}$.

Equation of the directrix

$y = -\dfrac{p}{2}$,

Coordinates of the focus

$F\left(0, \dfrac{p}{2}\right)$,

Coordinates of the vertex
$M(0, 0)$.

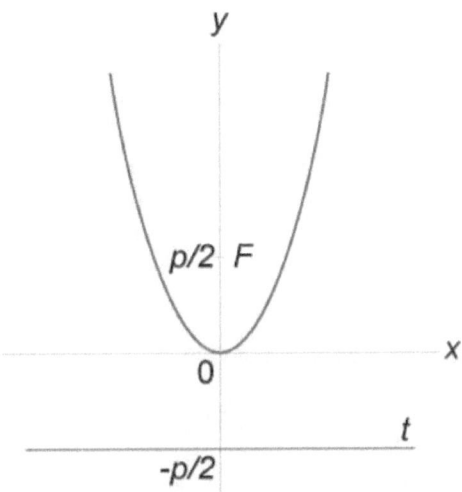

Figure 121.

669. **General Form, Axis Parallel to the y-axis**
$Ax^2 + Dx + Ey + F = 0$ (A, E nonzero),

$y = ax^2 + bx + c$, $p = \dfrac{1}{2a}$.

Equation of the directrix

$y = y_0 - \dfrac{p}{2}$,

Coordinates of the focus

$$F\left(x_0, y_0 + \frac{p}{2}\right),$$

Coordinates of the vertex

$$x_0 = -\frac{b}{2a}, \quad y_0 = ax_0^2 + bx_0 + c = \frac{4ac - b^2}{4a}.$$

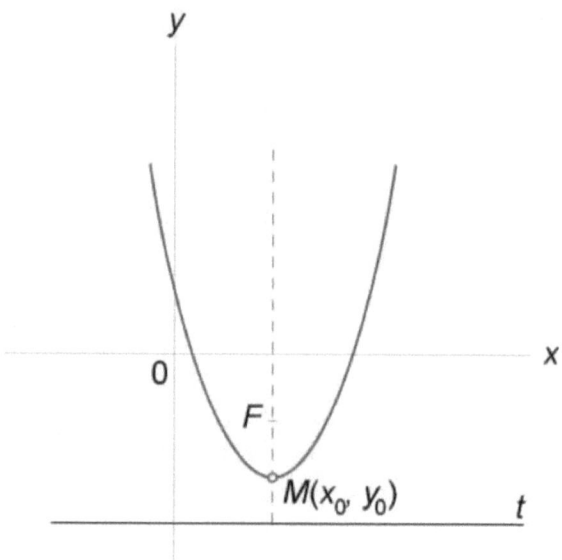

Figure 122.

7.8 Three-Dimensional Coordinate System

Point coordinates: x_0, y_0, z_0, x_1, y_1, z_1, ...
Real number: λ
Distance between two points: d
Area: S
Volume: V

670. Distance Between Two Points

$$d = AB = \sqrt{(x_2 - x_1)^2 + (y_2 - y_1)^2 + (z_2 - z_1)^2}$$

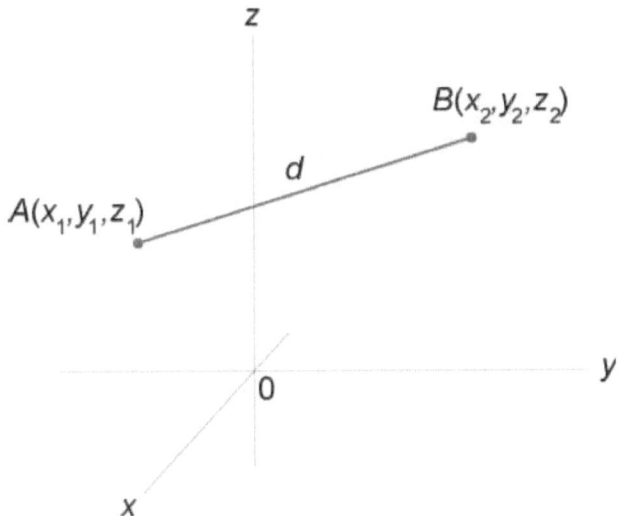

Figure 123.

671. Dividing a Line Segment in the Ratio λ

$$x_0 = \frac{x_1 + \lambda x_2}{1 + \lambda}, \quad y_0 = \frac{y_1 + \lambda y_2}{1 + \lambda}, \quad z_0 = \frac{z_1 + \lambda z_2}{1 + \lambda},$$

where

$$\lambda = \frac{AC}{CB}, \quad \lambda \neq -1.$$

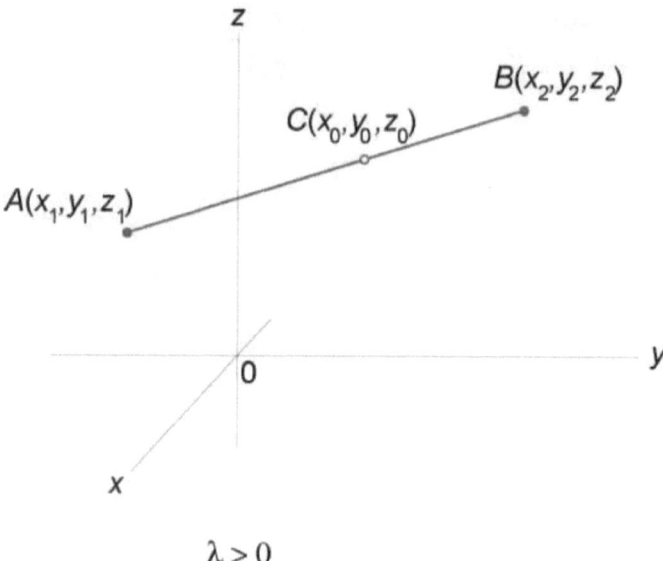

$\lambda > 0$

Figure 124.

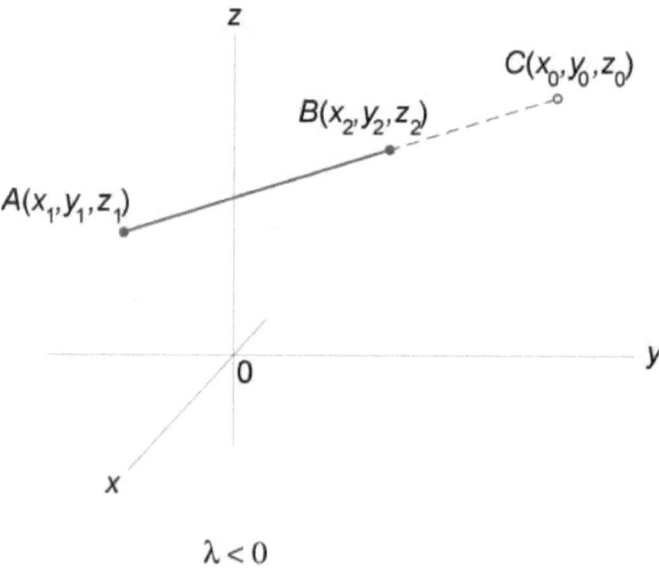

$\lambda < 0$

Figure 125.

CHAPTER 7. ANALYTIC GEOMETRY

672. Midpoint of a Line Segment
$$x_0 = \frac{x_1 + x_2}{2}, \quad y_0 = \frac{y_1 + y_2}{2}, \quad z_0 = \frac{z_1 + z_2}{2}, \quad \lambda = 1.$$

673. Area of a Triangle
The area of a triangle with vertices $P_1(x_1, y_1, z_1)$, $P_2(x_2, y_2, z_2)$, and $P_3(x_3, y_3, z_3)$ is given by
$$S = \frac{1}{2}\sqrt{\begin{vmatrix} y_1 & z_1 & 1 \\ y_2 & z_2 & 1 \\ y_3 & z_3 & 1 \end{vmatrix}^2 + \begin{vmatrix} z_1 & x_1 & 1 \\ z_2 & x_2 & 1 \\ z_3 & x_3 & 1 \end{vmatrix}^2 + \begin{vmatrix} x_1 & y_1 & 1 \\ x_2 & y_2 & 1 \\ x_3 & y_3 & 1 \end{vmatrix}^2}.$$

674. Volume of a Tetrahedron
The volume of a tetrahedron with vertices $P_1(x_1, y_1, z_1)$, $P_2(x_2, y_2, z_2)$, $P_3(x_3, y_3, z_3)$, and $P_4(x_4, y_4, z_4)$ is given by
$$V = \pm \frac{1}{6} \begin{vmatrix} x_1 & y_1 & z_1 & 1 \\ x_2 & y_2 & z_2 & 1 \\ x_3 & y_3 & z_3 & 1 \\ x_4 & y_4 & z_4 & 1 \end{vmatrix},$$
or
$$V = \pm \frac{1}{6} \begin{vmatrix} x_1 - x_4 & y_1 - y_4 & z_1 - z_4 \\ x_2 - x_4 & y_2 - y_4 & z_2 - z_4 \\ x_3 - x_4 & y_3 - y_4 & z_3 - z_4 \end{vmatrix}.$$

Note: We choose the sign (+) or (−) so that to get a positive answer for volume.

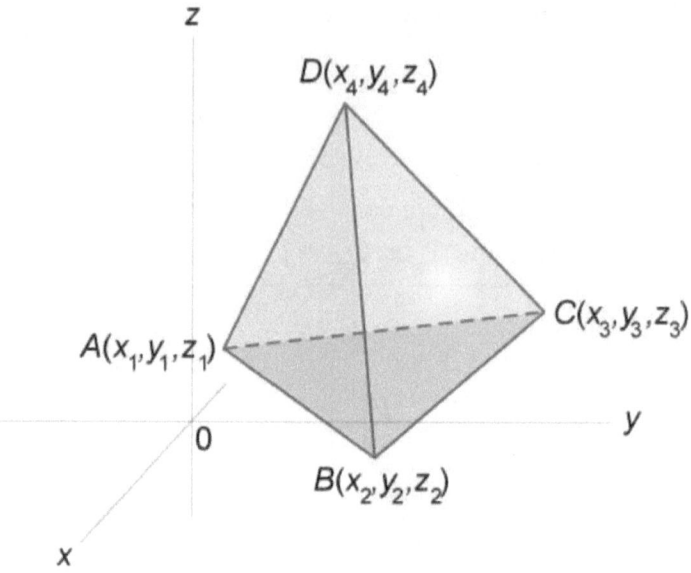

Figure 126.

7.9 Plane

Point coordinates: x, y, z, x_0, y_0, z_0, x_1, y_1, z_1, ...
Real numbers: A, B, C, D, A_1, A_2, a, b, c, a_1, a_2, λ, p, t, ...
Normal vectors: $\vec{n}$, $\vec{n}_1$, $\vec{n}_2$
Direction cosines: $\cos\alpha$, $\cos\beta$, $\cos\gamma$
Distance from point to plane: d

675. General Equation of a Plane
$Ax + By + Cz + D = 0$

676. Normal Vector to a Plane

The vector $\bar{n}(A, B, C)$ is normal to the plane $Ax + By + Cz + D = 0$.

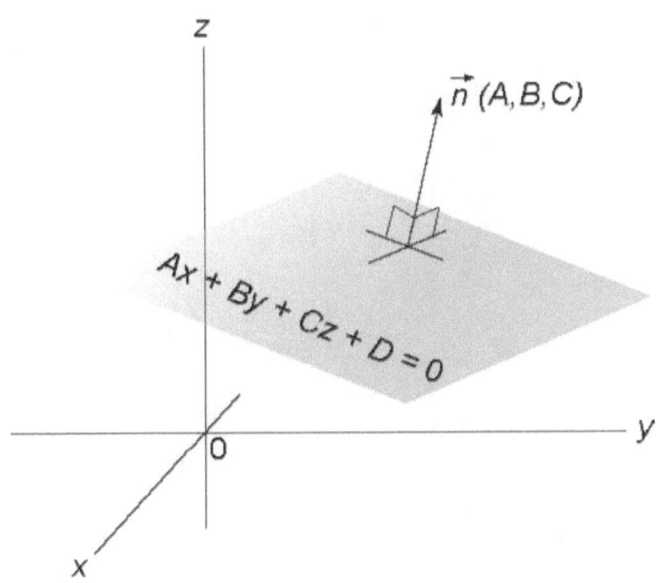

Figure 127.

677. Particular Cases of the Equation of a Plane
$Ax + By + Cz + D = 0$

If $A = 0$, the plane is parallel to the x-axis.
If $B = 0$, the plane is parallel to the y-axis.
If $C = 0$, the plane is parallel to the z-axis.
If $D = 0$, the plane lies on the origin.

If $A = B = 0$, the plane is parallel to the xy-plane.
If $B = C = 0$, the plane is parallel to the yz-plane.
If $A = C = 0$, the plane is parallel to the xz-plane.

CHAPTER 7. ANALYTIC GEOMETRY

678. Point Direction Form
$$A(x-x_0)+B(y-y_0)+C(z-z_0)=0,$$
where the point $P(x_0,y_0,z_0)$ lies in the plane, and the vector (A, B, C) is normal to the plane.

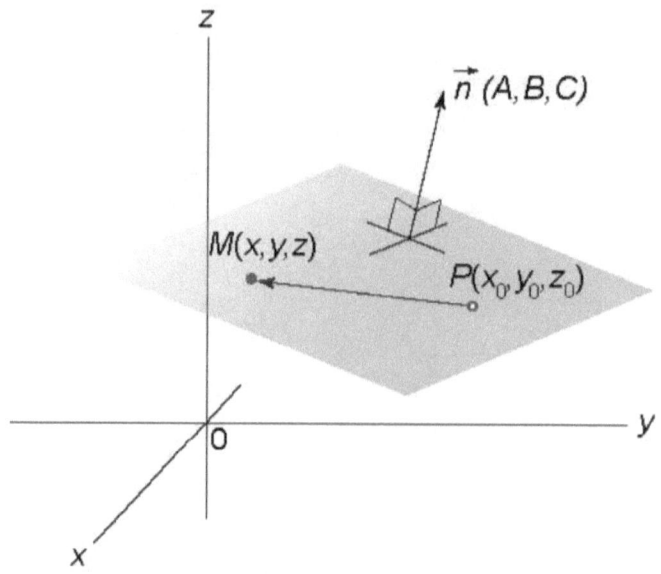

Figure 128.

679. Intercept Form
$$\frac{x}{a}+\frac{y}{b}+\frac{z}{c}=1$$

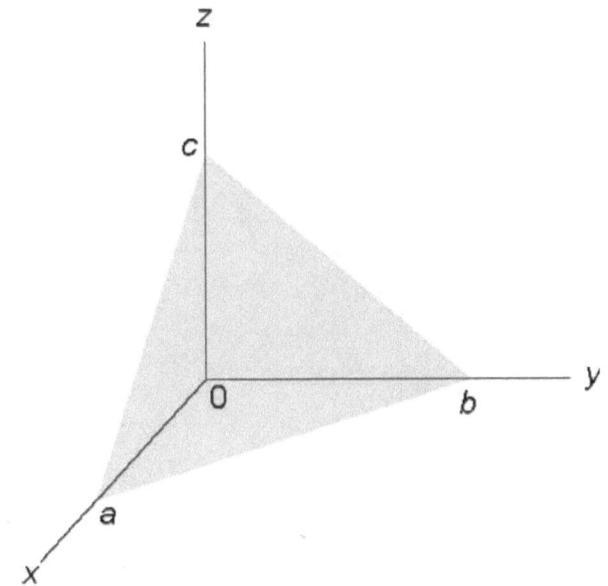

Figure 129.

680. Three Point Form
$$\begin{vmatrix} x-x_3 & y-y_3 & z-z_3 \\ x_1-x_3 & y_1-y_3 & z_1-z_3 \\ x_2-x_3 & y_2-y_3 & z_2-z_3 \end{vmatrix} = 0,$$
or
$$\begin{vmatrix} x & y & z & 1 \\ x_1 & y_1 & z_1 & 1 \\ x_2 & y_2 & z_2 & 1 \\ x_3 & y_3 & z_3 & 1 \end{vmatrix} = 0.$$

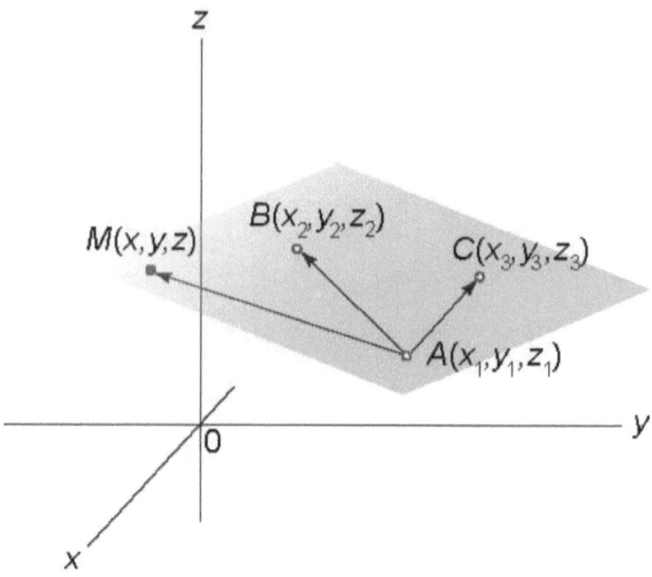

Figure 130.

681. Normal Form
$$x\cos\alpha + y\cos\beta + z\cos\gamma - p = 0,$$

where p is the perpendicular distance from the origin to the plane, and $\cos\alpha$, $\cos\beta$, $\cos\gamma$ are the direction cosines of any line normal to the plane.

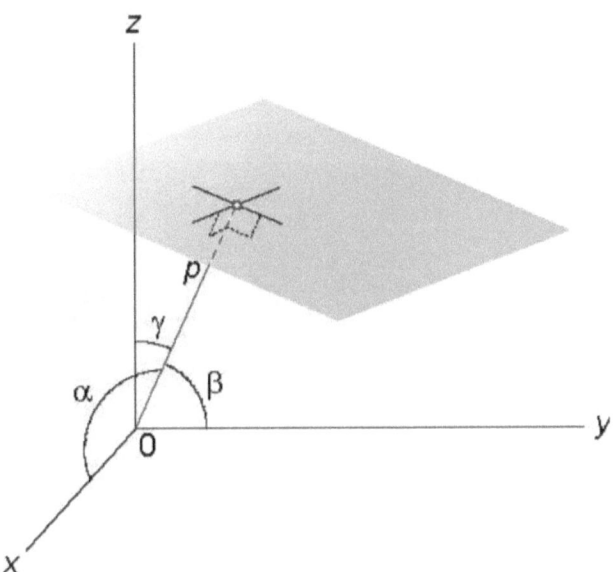

Figure 131.

682. Parametric Form

$$\begin{cases} x = x_1 + a_1 s + a_2 t \\ y = y_1 + b_1 s + b_2 t, \\ z = z_1 + c_1 s + c_2 t \end{cases}$$

where (x, y, z) are the coordinates of any unknown point on the line, the point $P(x_1, y_1, z_1)$ lies in the plane, the vectors (a_1, b_1, c_1) and (a_2, b_2, c_2) are parallel to the plane.

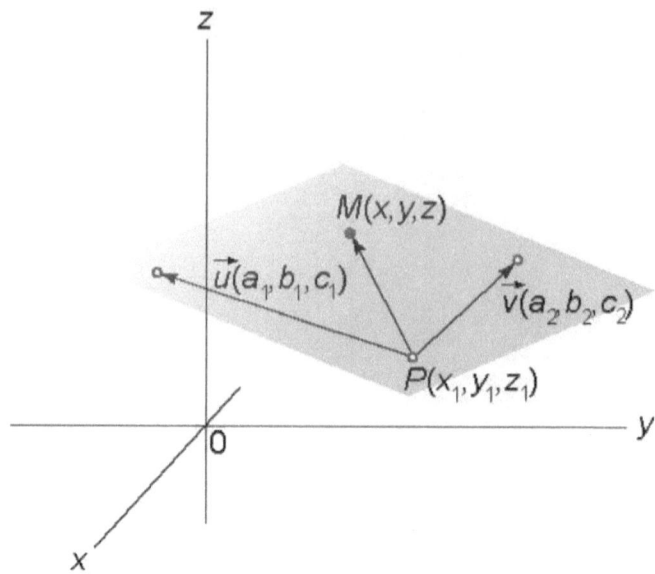

Figure 132.

683. Dihedral Angle Between Two Planes
If the planes are given by
$A_1 x + B_1 y + C_1 z + D_1 = 0$,
$A_2 x + B_2 y + C_2 z + D_2 = 0$,
then the dihedral angle between them is
$$\cos \varphi = \frac{\vec{n}_1 \cdot \vec{n}_2}{|\vec{n}_1| \cdot |\vec{n}_2|} = \frac{A_1 A_2 + B_1 B_2 + C_1 C_2}{\sqrt{A_1^2 + B_1^2 + C_1^2} \cdot \sqrt{A_2^2 + B_2^2 + C_2^2}}.$$

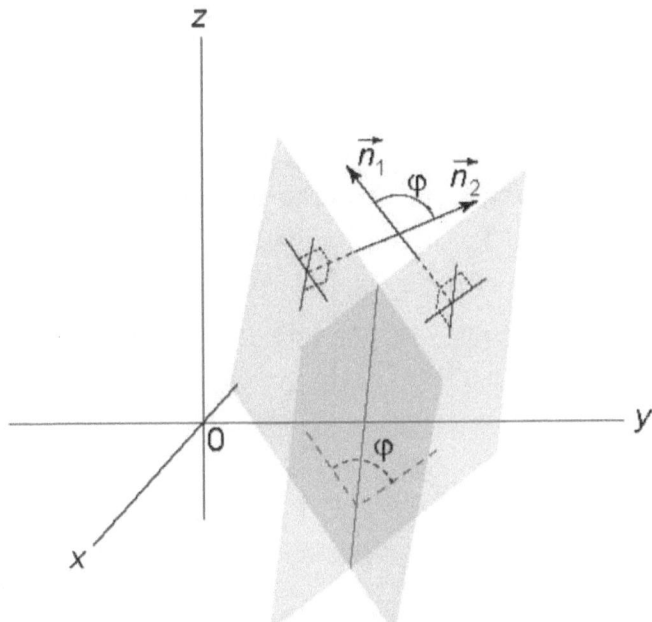

Figure 133.

684. Parallel Planes
Two planes $A_1 x + B_1 y + C_1 z + D_1 = 0$ and $A_2 x + B_2 y + C_2 z + D_2 = 0$ are parallel if
$$\frac{A_1}{A_2} = \frac{B_1}{B_2} = \frac{C_1}{C_2}.$$

685. Perpendicular Planes
Two planes $A_1 x + B_1 y + C_1 z + D_1 = 0$ and $A_2 x + B_2 y + C_2 z + D_2 = 0$ are perpendicular if
$A_1 A_2 + B_1 B_2 + C_1 C_2 = 0$.

686. Equation of a Plane Through $P(x_1, y_1, z_1)$ and Parallel To the Vectors (a_1, b_1, c_1) and (a_2, b_2, c_2) (Fig.132)

$$\begin{vmatrix} x-x_1 & y-y_1 & z-z_1 \\ a_1 & b_1 & c_1 \\ a_2 & b_2 & c_2 \end{vmatrix} = 0$$

687. Equation of a Plane Through $P_1(x_1,y_1,z_1)$ and $P_2(x_2,y_2,z_2)$, and Parallel To the Vector (a, b, c)

$$\begin{vmatrix} x-x_1 & y-y_1 & z-z_1 \\ x_2-x_1 & y_2-y_1 & z_2-z_1 \\ a & b & c \end{vmatrix} = 0$$

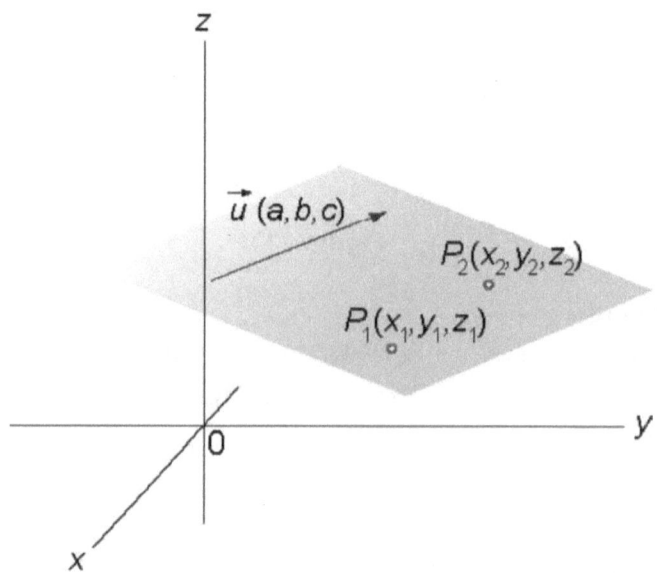

Figure 134.

688. Distance From a Point To a Plane

The distance from the point $P_1(x_1,y_1,z_1)$ to the plane $Ax+By+Cz+D=0$ is

$$d = \left| \frac{Ax_1 + By_1 + Cz_1 + D}{\sqrt{A^2 + B^2 + C^2}} \right|.$$

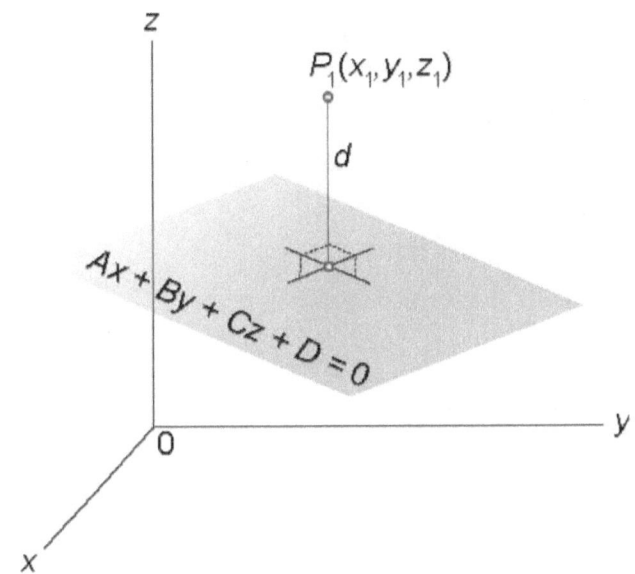

Figure 135.

689. Intersection of Two Planes

If two planes $A_1 x + B_1 y + C_1 z + D_1 = 0$ and $A_2 x + B_2 y + C_2 z + D_2 = 0$ intersect, the intersection straight line is given by
$$\begin{cases} x = x_1 + at \\ y = y_1 + bt, \\ z = z_1 + ct \end{cases}$$
or
$$\frac{x - x_1}{a} = \frac{y - y_1}{b} = \frac{z - z_1}{c},$$
where

$$a = \begin{vmatrix} B_1 & C_1 \\ B_2 & C_2 \end{vmatrix}, \quad b = \begin{vmatrix} C_1 & A_1 \\ C_2 & A_2 \end{vmatrix}, \quad c = \begin{vmatrix} A_1 & B_1 \\ A_2 & B_2 \end{vmatrix},$$

$$x_1 = \frac{b\begin{vmatrix} D_1 & C_1 \\ D_2 & C_2 \end{vmatrix} - c\begin{vmatrix} D_1 & B_1 \\ D_2 & B_2 \end{vmatrix}}{a^2 + b^2 + c^2},$$

$$y_1 = \frac{c\begin{vmatrix} D_1 & A_1 \\ D_2 & A_2 \end{vmatrix} - a\begin{vmatrix} D_1 & C_1 \\ D_2 & C_2 \end{vmatrix}}{a^2 + b^2 + c^2},$$

$$z_1 = \frac{a\begin{vmatrix} D_1 & B_1 \\ D_2 & B_2 \end{vmatrix} - b\begin{vmatrix} D_1 & A_1 \\ D_2 & A_2 \end{vmatrix}}{a^2 + b^2 + c^2}.$$

7.10 Straight Line in Space

Point coordinates: $x, y, z, x_1, y_1, z_1, \ldots$
Direction cosines: $\cos\alpha, \cos\beta, \cos\gamma$
Real numbers: $A, B, C, D, a, b, c, a_1, a_2, t, \ldots$
Direction vectors of a line: $\vec{s}, \vec{s}_1, \vec{s}_2$
Normal vector to a plane: $\vec{n}$
Angle between two lines: φ

690. Point Direction Form of the Equation of a Line
$$\frac{x - x_1}{a} = \frac{y - y_1}{b} = \frac{z - z_1}{c},$$
where the point $P_1(x_1, y_1, z_1)$ lies on the line, and (a, b, c) is the direction vector of the line.

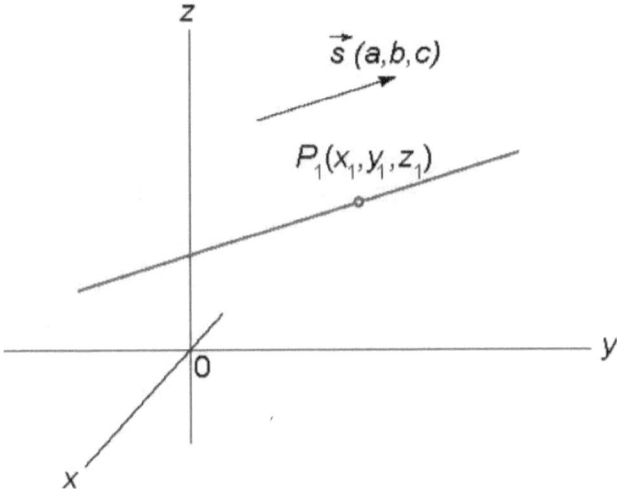

Figure 136.

691. Two Point Form

$$\frac{x-x_1}{x_2-x_1} = \frac{y-y_1}{y_2-y_1} = \frac{z-z_1}{z_2-z_1}$$

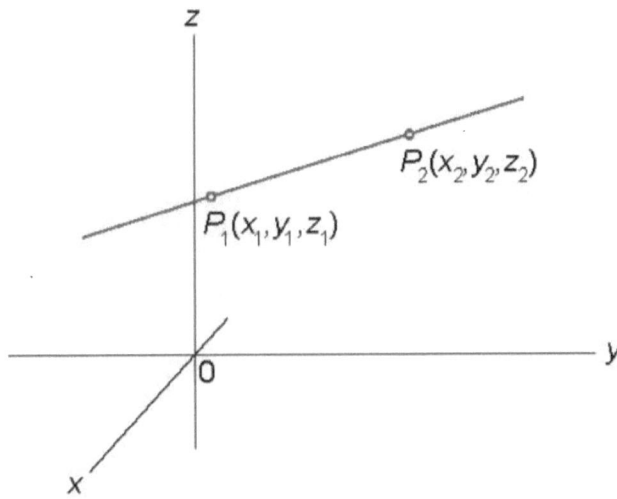

Figure 137.

692. Parametric Form

$$\begin{cases} x = x_1 + t\cos\alpha \\ y = y_1 + t\cos\beta \\ z = z_1 + t\cos\gamma \end{cases},$$

where the point $P_1(x_1, y_1, z_1)$ lies on the straight line, $\cos\alpha$, $\cos\beta$, $\cos\gamma$ are the direction cosines of the direction vector of the line, the parameter t is any real number.

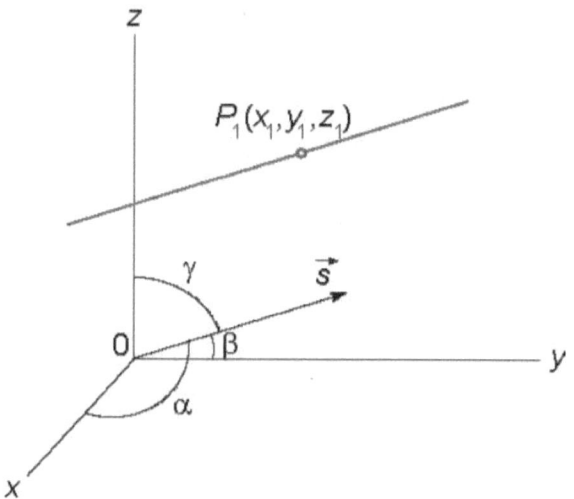

Figure 138.

693. Angle Between Two Straight Lines

$$\cos\varphi = \frac{\vec{s}_1 \cdot \vec{s}_2}{|\vec{s}_1| \cdot |\vec{s}_2|} = \frac{a_1 a_2 + b_1 b_2 + c_1 c_2}{\sqrt{a_1^2 + b_1^2 + c_1^2} \cdot \sqrt{a_2^2 + b_2^2 + c_2^2}}$$

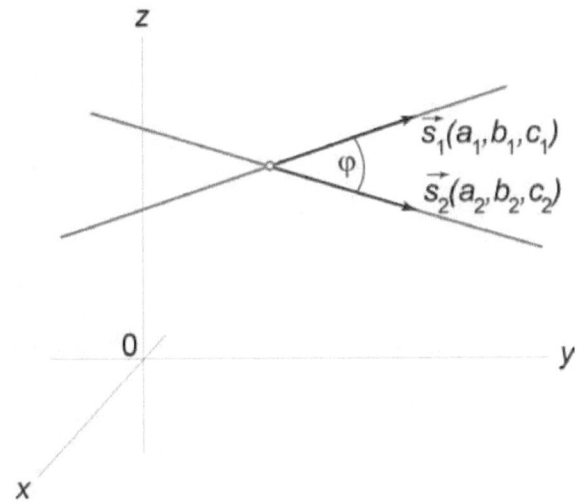

Figure 139.

694. Parallel Lines
Two lines are parallel if
$\vec{s}_1 \parallel \vec{s}_2$,
or
$$\frac{a_1}{a_2} = \frac{b_1}{b_2} = \frac{c_1}{c_2}.$$

695. Perpendicular Lines
Two lines are parallel if
$\vec{s}_1 \cdot \vec{s}_2 = 0$,
or
$a_1 a_2 + b_1 b_2 + c_1 c_2 = 0$.

696. Intersection of Two Lines
Two lines $\dfrac{x - x_1}{a_1} = \dfrac{y - y_1}{b_1} = \dfrac{z - z_1}{c_1}$ and

$$\frac{x-x_2}{a_2} = \frac{y-y_2}{b_2} = \frac{z-z_2}{c_2} \text{ intersect if}$$

$$\begin{vmatrix} x_2-x_1 & y_2-y_1 & z_2-z_1 \\ a_1 & b_1 & c_1 \\ a_2 & b_2 & c_2 \end{vmatrix} = 0.$$

697. Parallel Line and Plane

The straight line $\dfrac{x-x_1}{a} = \dfrac{y-y_1}{b} = \dfrac{z-z_1}{c}$ and the plane $Ax+By+Cz+D=0$ are parallel if
$\vec{n} \cdot \vec{s} = 0$,
or
$Aa+Bb+Cc=0$.

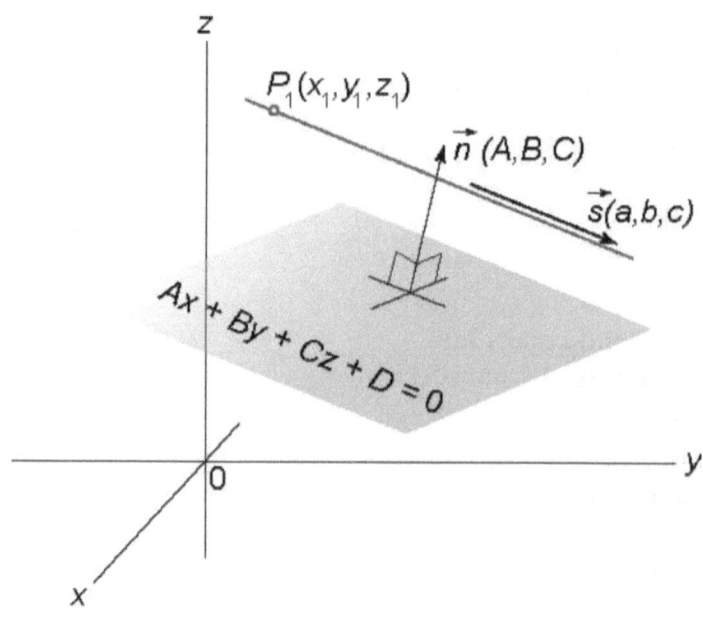

Figure 140.

698. Perpendicular Line and Plane

The straight line $\dfrac{x-x_1}{a} = \dfrac{y-y_1}{b} = \dfrac{z-z_1}{c}$ and the plane $Ax + By + Cz + D = 0$ are perpendicular if $\vec{n} \| \vec{s}$,

or

$$\dfrac{A}{a} = \dfrac{B}{b} = \dfrac{C}{c}.$$

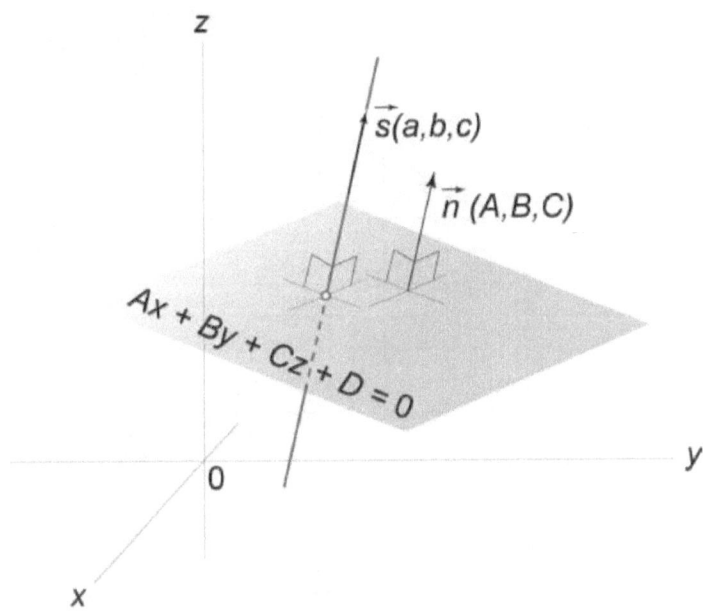

Figure 141.

7.11 Quadric Surfaces

Point coordinates of the quadric surfaces: x, y, z
Real numbers: A, B, C, a, b, c, $k_1, k_2, k_3, \ldots$

699. General Quadratic Equation

$$Ax^2 + By^2 + Cz^2 + 2Fyz + 2Gzx + 2Hxy + 2Px + 2Qy + 2Rz + D = 0$$

700. Classification of Quadric Surfaces

Case	Rank(e)	Rank(E)	Δ	k signs	Type of Surface
1	3	4	<0	Same	Real Ellipsoid
2	3	4	>0	Same	Imaginary Ellipsoid
3	3	4	>0	Different	Hyperboloid of 1 Sheet
4	3	4	<0	Different	Hyperboloid of 2 Sheets
5	3	3		Different	Real Quadric Cone
6	3	3		Same	Imaginary Quadric Cone
7	2	4	<0	Same	Elliptic Paraboloid
8	2	4	>0	Different	Hyperbolic Paraboloid
9	2	3		Same	Real Elliptic Cylinder
10	2	3		Same	Imaginary Elliptic Cylinder
11	2	3		Different	Hyperbolic Cylinder
12	2	2		Different	Real Intersecting Planes
13	2	2		Same	Imaginary Intersecting Planes
14	1	3			Parabolic Cylinder
15	1	2			Real Parallel Planes
16	1	2			Imaginary Parallel Planes
17	1	1			Coincident Planes

Here

$$e = \begin{pmatrix} A & H & G \\ H & B & F \\ G & F & C \end{pmatrix}, \quad E = \begin{pmatrix} A & H & Q & P \\ H & B & F & Q \\ G & F & C & R \\ P & Q & R & D \end{pmatrix}, \quad \Delta = \det(E),$$

k_1, k_2, k_3 are the roots of the equation,

$$\begin{vmatrix} A-x & H & G \\ H & B-x & F \\ G & F & C-x \end{vmatrix} = 0.$$

CHAPTER 7. ANALYTIC GEOMETRY

701. Real Ellipsoid (Case 1)
$$\frac{x^2}{a^2}+\frac{y^2}{b^2}+\frac{z^2}{c^2}=1$$

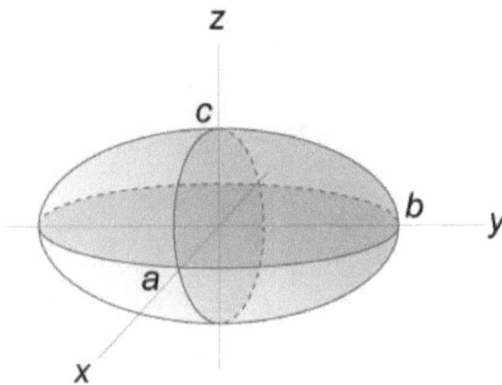

Figure 142.

702. Imaginary Ellipsoid (Case 2)
$$\frac{x^2}{a^2}+\frac{y^2}{b^2}+\frac{z^2}{c^2}=-1$$

703. Hyperboloid of 1 Sheet (Case 3)
$$\frac{x^2}{a^2}+\frac{y^2}{b^2}-\frac{z^2}{c^2}=1$$

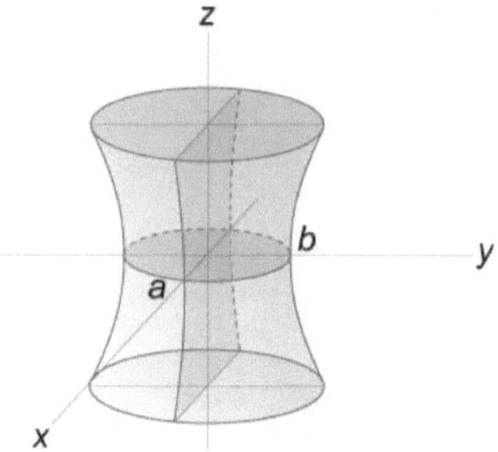

Figure 143.

704. Hyperboloid of 2 Sheets (Case 4)
$$\frac{x^2}{a^2}+\frac{y^2}{b^2}-\frac{z^2}{c^2}=-1$$

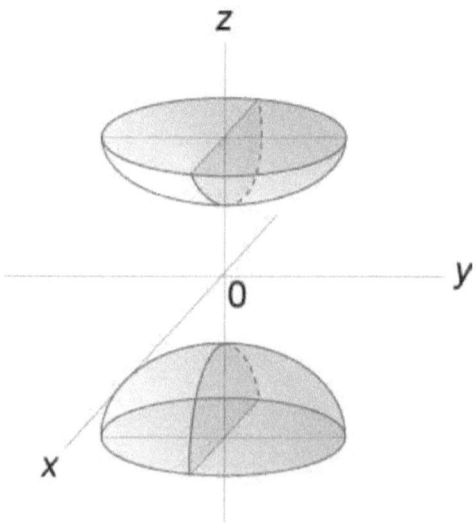

Figure 144.

705. Real Quadric Cone (Case 5)
$$\frac{x^2}{a^2}+\frac{y^2}{b^2}-\frac{z^2}{c^2}=0$$

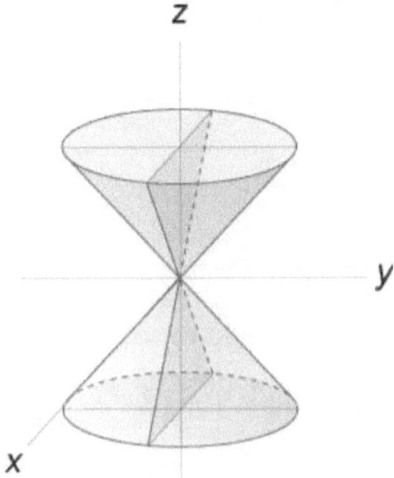

Figure 145.

706. Imaginary Quadric Cone (Case 6)
$$\frac{x^2}{a^2}+\frac{y^2}{b^2}+\frac{z^2}{c^2}=0$$

707. Elliptic Paraboloid (Case 7)
$$\frac{x^2}{a^2}+\frac{y^2}{b^2}-z=0$$

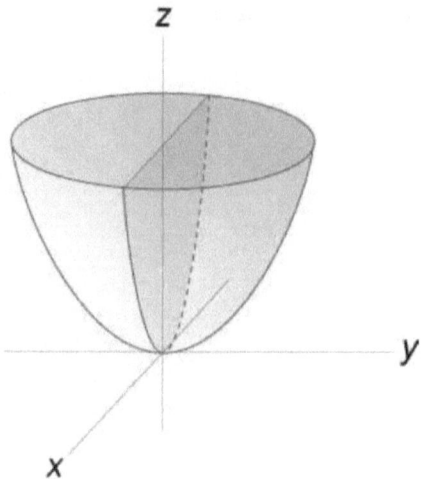

Figure 146.

708. Hyperbolic Paraboloid (Case 8)

$$\frac{x^2}{a^2} - \frac{y^2}{b^2} - z = 0$$

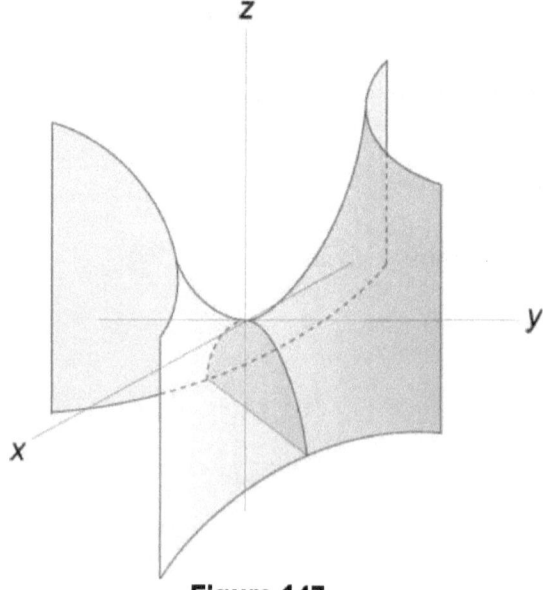

Figure 147.

Real Elliptic Cylinder (Case 9)
$$\frac{x^2}{a^2} + \frac{y^2}{b^2} = 1$$

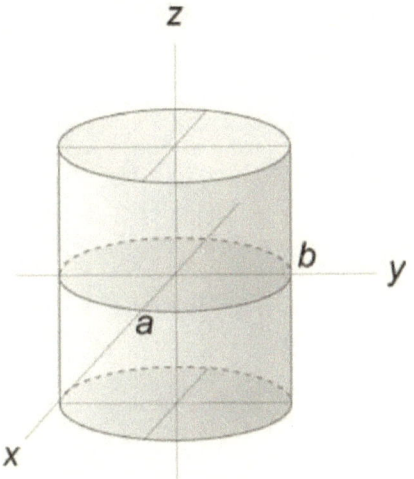

Figure 148.

710. Imaginary Elliptic Cylinder (Case 10)
$$\frac{x^2}{a^2} + \frac{y^2}{b^2} = -1$$

711. Hyperbolic Cylinder (Case 11)
$$\frac{x^2}{a^2} - \frac{y^2}{b^2} = 1$$

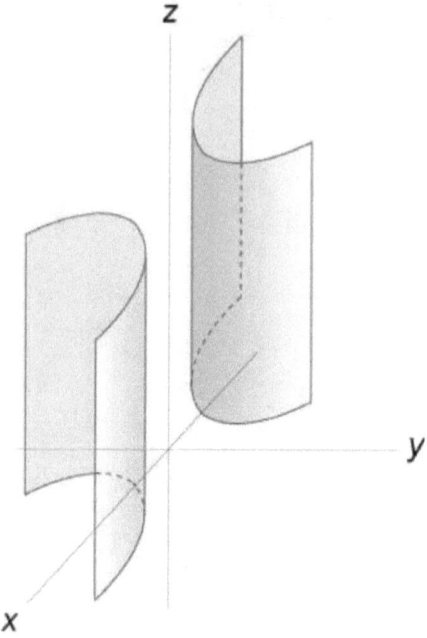

Figure 149.

712. Real Intersecting Planes (Case 12)

$$\frac{x^2}{a^2} - \frac{y^2}{b^2} = 0$$

713. Imaginary Intersecting Planes (Case 13)

$$\frac{x^2}{a^2} + \frac{y^2}{b^2} = 0$$

714. Parabolic Cylinder (Case 14)

$$\frac{x^2}{a^2} - y = 0$$

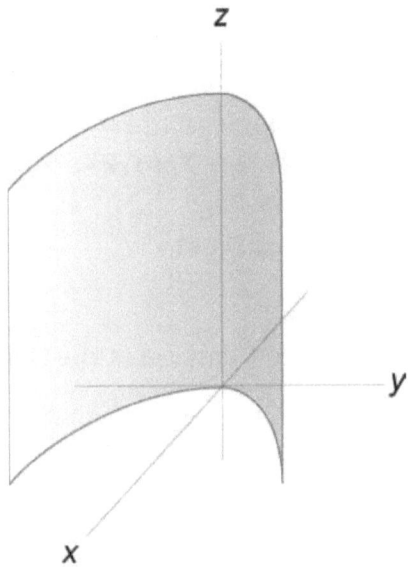

Figure 150.

715. Real Parallel Planes (Case 15)
$$\frac{x^2}{a^2} = 1$$

716. Imaginary Parallel Planes (Case 16)
$$\frac{x^2}{a^2} = -1$$

717. Coincident Planes (Case 17)
$$x^2 = 0$$

7.12 Sphere

Radius of a sphere: R
Point coordinates: x, y, z, x_1, y_1, z_1, ...
Center of a sphere: (a, b, c)
Real numbers: A, D, E, F, M

718. Equation of a Sphere Centered at the Origin (Standard Form)
$$x^2 + y^2 + z^2 = R^2$$

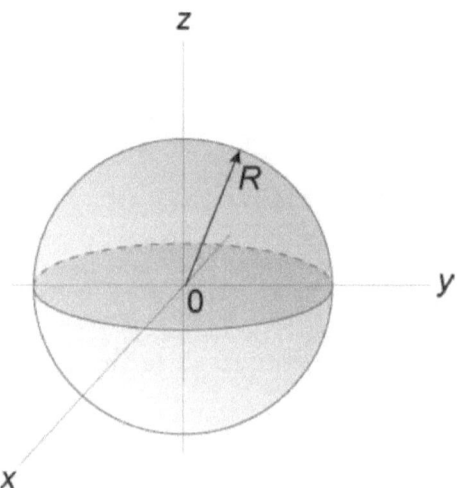

Figure 151.

719. Equation of a Circle Centered at Any Point (a, b, c)
$$(x-a)^2 + (y-b)^2 + (z-c)^2 = R^2$$

720. Diameter Form
$$(x-x_1)(x-x_2) + (y-y_1)(y-y_2) + (z-z_1)(z-z_2) = 0,$$

where
$P_1(x_1,y_1,z_1)$, $P_2(x_2,y_2,z_2)$ are the ends of a diameter.

721. Four Point Form

$$\begin{vmatrix} x^2+y^2+z^2 & x & y & z & 1 \\ x_1^2+y_1^2+x_1^2 & x_1 & y_1 & z_1 & 1 \\ x_2^2+y_2^2+x_2^2 & x_2 & y_2 & z_2 & 1 \\ x_3^2+y_3^2+x_3^2 & x_3 & y_3 & z_3 & 1 \\ x_4^2+y_4^2+x_4^2 & x_4 & y_4 & z_4 & 1 \end{vmatrix} = 0$$

722. General Form
$Ax^2 + Ay^2 + Az^2 + Dx + Ey + Fz + M = 0$ (A is nonzero).
The center of the sphere has coordinates (a,b,c), where
$$a = -\frac{D}{2A}, \quad b = -\frac{E}{2A}, \quad c = -\frac{F}{2A}.$$
The radius of the sphere is
$$R = \frac{\sqrt{D^2 + E^2 + F^2 - 4A^2M}}{2A}.$$

Chapter 8
Differential Calculus

Functions: f, g, y, u, v
Argument (independent variable): x
Real numbers: a, b, c, d
Natural number: n
Angle: α
Inverse function: f^{-1}

8.1 Functions and Their Graphs

723. Even Function
$f(-x) = f(x)$

724. Odd Function
$f(-x) = -f(x)$

725. Periodic Function
$f(x + nT) = f(x)$

726. Inverse Function
$y = f(x)$ is any function, $x = g(y)$ or $y = f^{-1}(x)$ is its inverse function.

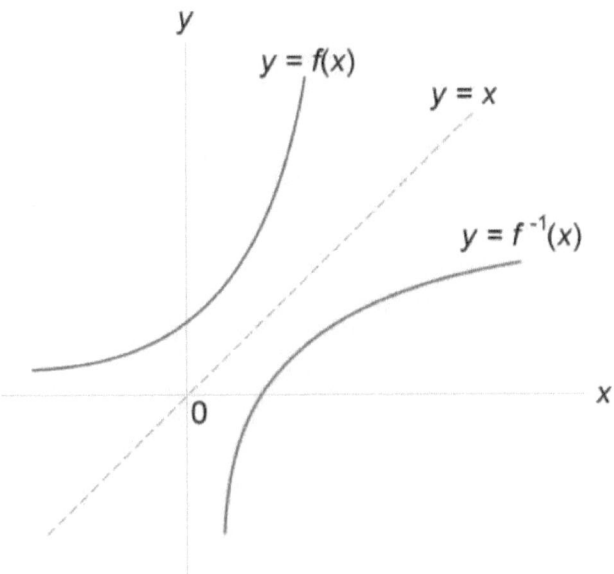

Figure 152.

727. Composite Function
$y = f(u)$, $u = g(x)$, $y = f(g(x))$ is a composite function.

728. Linear Function
$y = ax + b$, $x \in R$, $a = \tan \alpha$ is the slope of the line, b is the y-intercept.

CHAPTER 8. DIFFERENTIAL CALCULUS

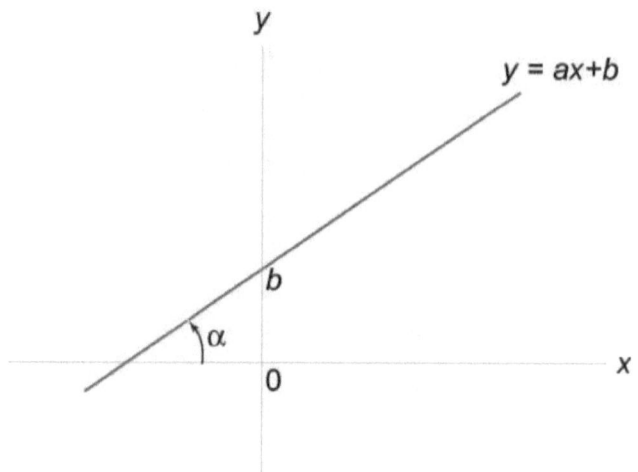

Figure 153.

729. Quadratic Function
$y = x^2$, $x \in R$.

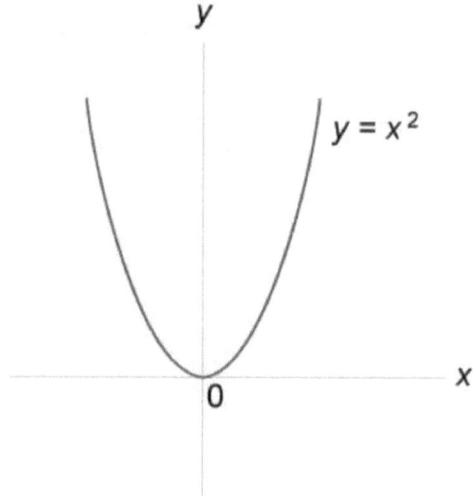

Figure 154.

CHAPTER 8. DIFFERENTIAL CALCULUS

730. $y = ax^2 + bx + c$, $x \in R$.

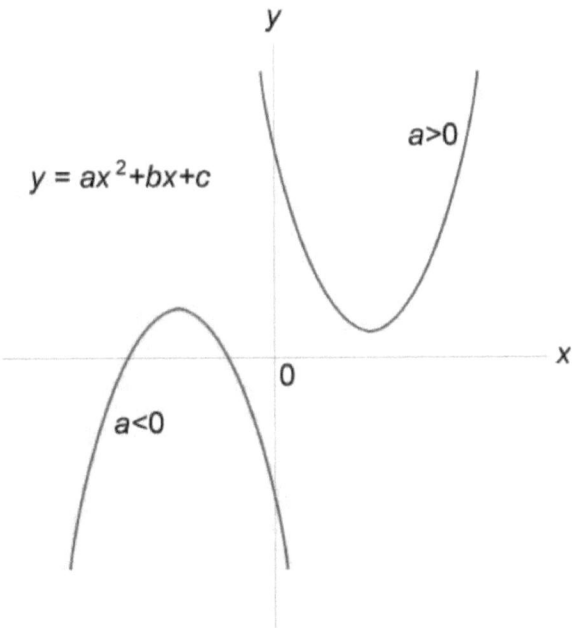

Figure 155.

731. Cubic Function
$y = x^3$, $x \in R$.

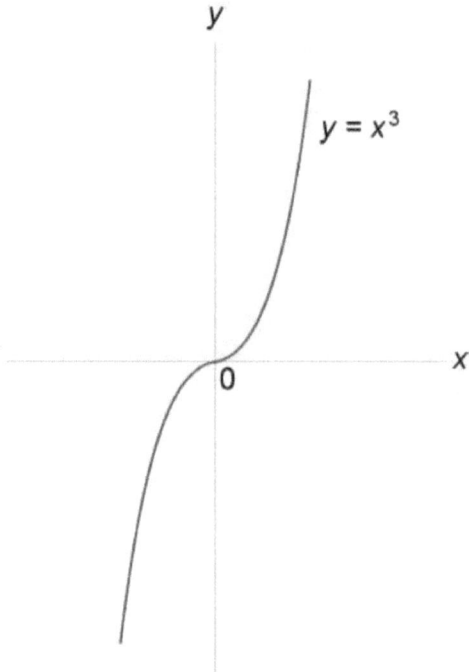

Figure 156.

732. $y = ax^3 + bx^2 + cx + d$, $x \in \mathbb{R}$.

CHAPTER 8. DIFFERENTIAL CALCULUS

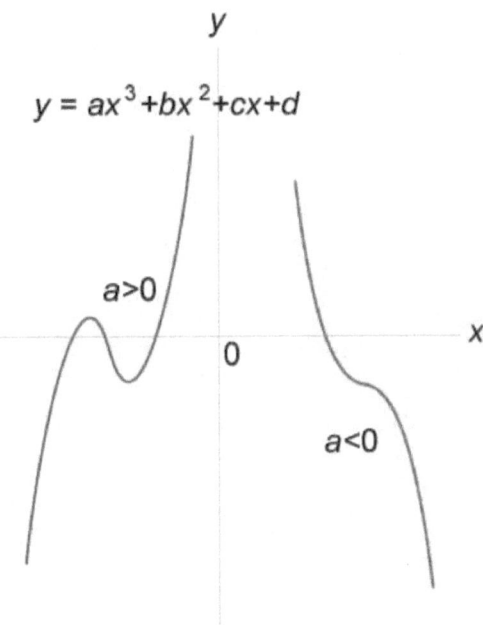

Figure 157.

733. Power Function
$y = x^n$, $n \in N$.

CHAPTER 8. DIFFERENTIAL CALCULUS

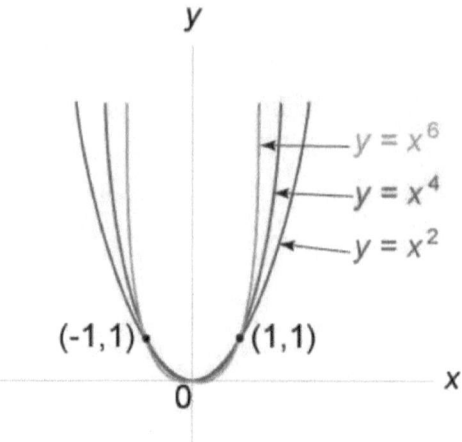

Figure 158.

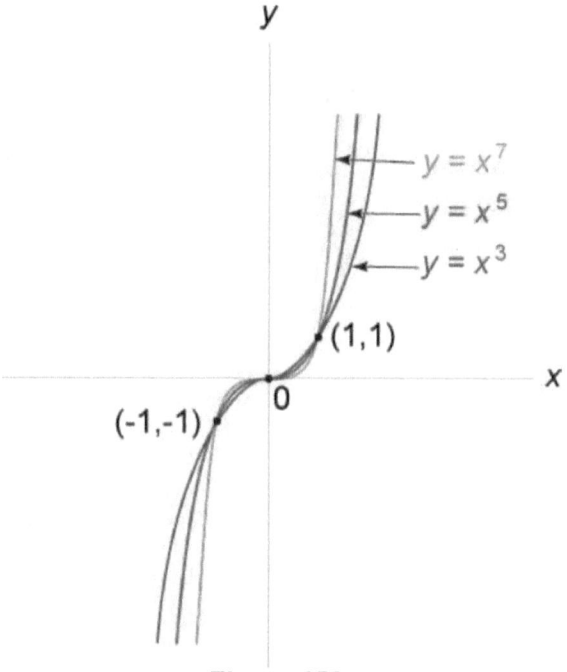

Figure 159.

734. Square Root Function
$y = \sqrt{x}$, $x \in [0, \infty)$.

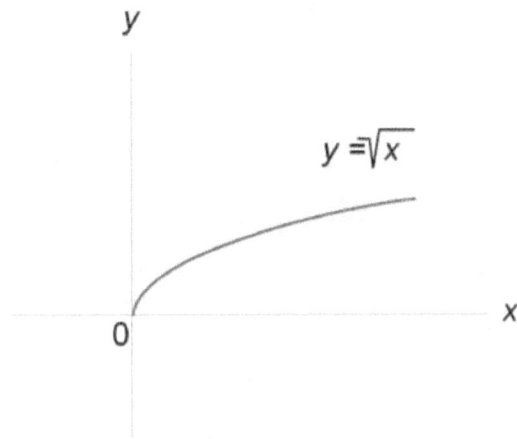

Figure 160.

735. Exponential Functions
$y = a^x$, $a > 0$, $a \neq 1$,
$y = e^x$ if $a = e$, $e = 2.71828182846...$

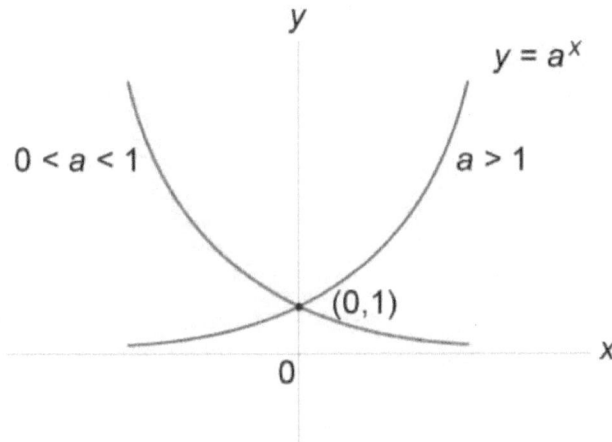

Figure 161.

736. Logarithmic Functions
$y = \log_a x$, $x \in (0, \infty)$, $a > 0$, $a \neq 1$,
$y = \ln x$ if $a = e$, $x > 0$.

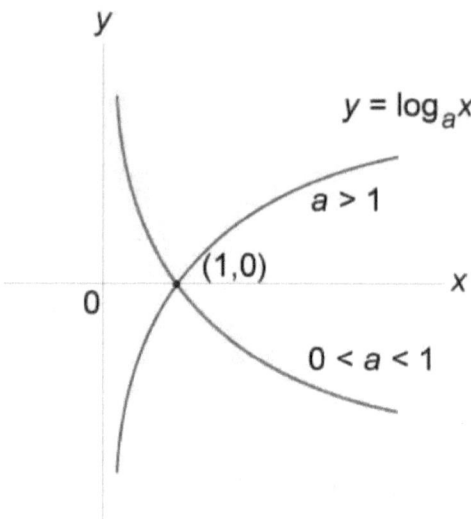

Figure 162.

737. Hyperbolic Sine Function
$y = \sinh x$, $\sinh x = \dfrac{e^x - e^{-x}}{2}$, $x \in R$.

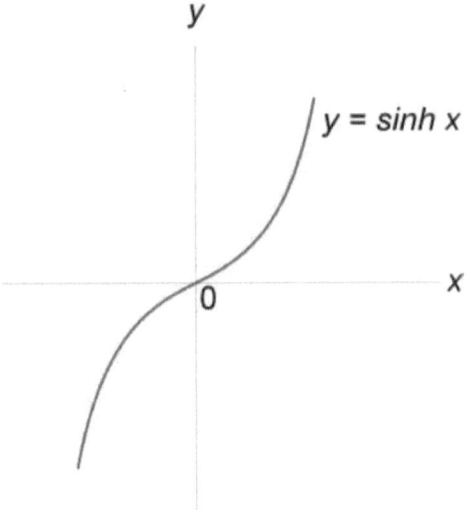

Figure 163.

738. Hyperbolic Cosine Function

$$y = \cosh x, \quad \cosh x = \frac{e^x + e^{-x}}{2}, \quad x \in \mathbb{R}.$$

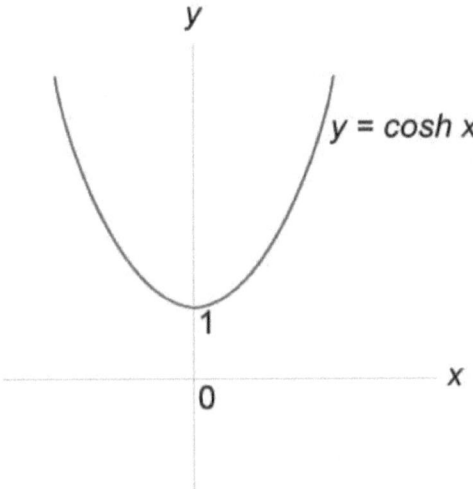

Figure 164.

739. Hyperbolic Tangent Function

$$y = \tanh x, \quad y = \tanh x = \frac{\sinh x}{\cosh x} = \frac{e^x - e^{-x}}{e^x + e^{-x}}, \quad x \in \mathbb{R}.$$

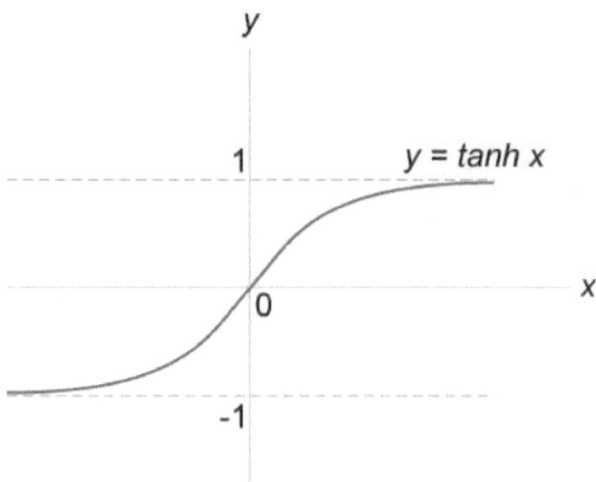

Figure 165.

740. Hyperbolic Cotangent Function

$$y = \coth x, \quad y = \coth x = \frac{\cosh x}{\sinh x} = \frac{e^x + e^{-x}}{e^x - e^{-x}}, \quad x \in \mathbb{R}, \ x \neq 0.$$

CHAPTER 8. DIFFERENTIAL CALCULUS

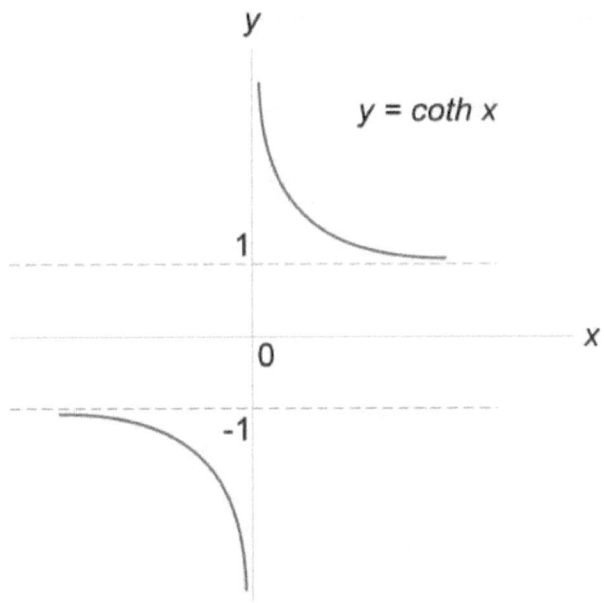

Figure 166.

741. Hyperbolic Secant Function

$$y = \operatorname{sech} x, \quad y = \operatorname{sech} x = \frac{1}{\cosh x} = \frac{2}{e^x + e^{-x}}, \quad x \in \mathbf{R}.$$

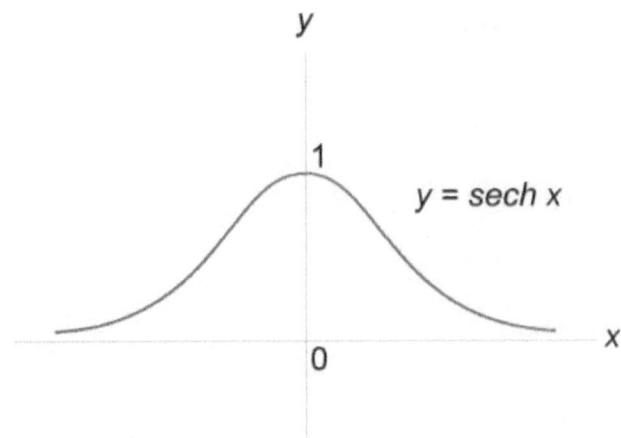

Figure 167.

742. Hyperbolic Cosecant Function

$$y = \operatorname{csch} x, \quad y = \operatorname{csch} x = \frac{1}{\sinh x} = \frac{2}{e^x - e^{-x}}, \quad x \in R, \ x \neq 0.$$

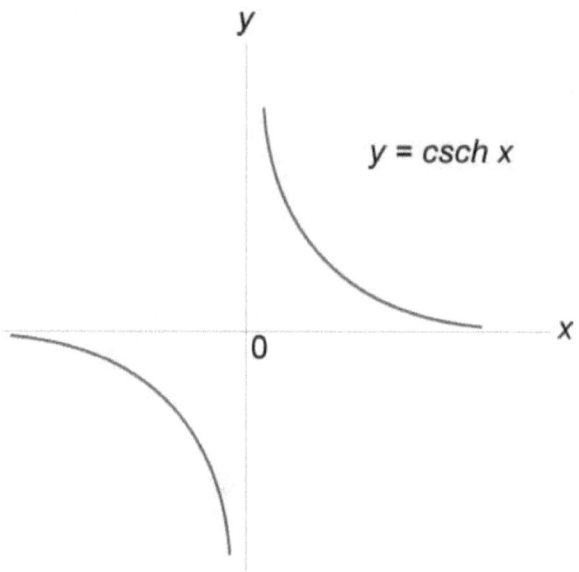

Figure 168.

743. Inverse Hyperbolic Sine Function

$y = \operatorname{arcsinh} x, \ x \in R$.

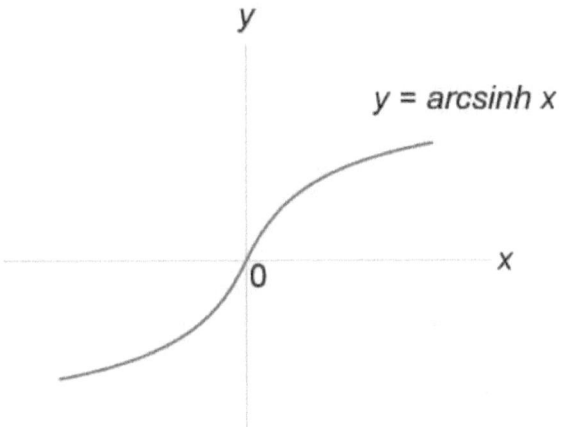

Figure 169.

744. Inverse Hyperbolic Cosine Function
$y = \text{arccosh } x$, $x \in [1, \infty)$.

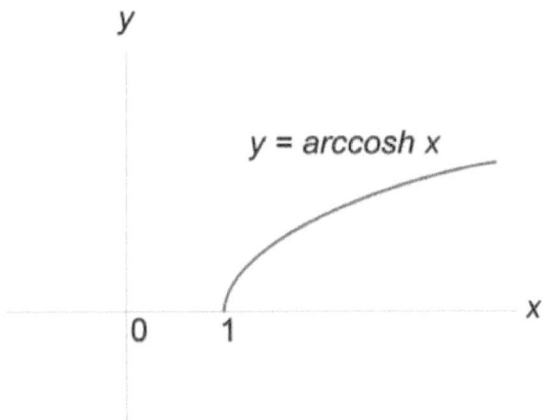

Figure 170.

745. Inverse Hyperbolic Tangent Function
$y = \text{arctanh } x$, $x \in (-1, 1)$.

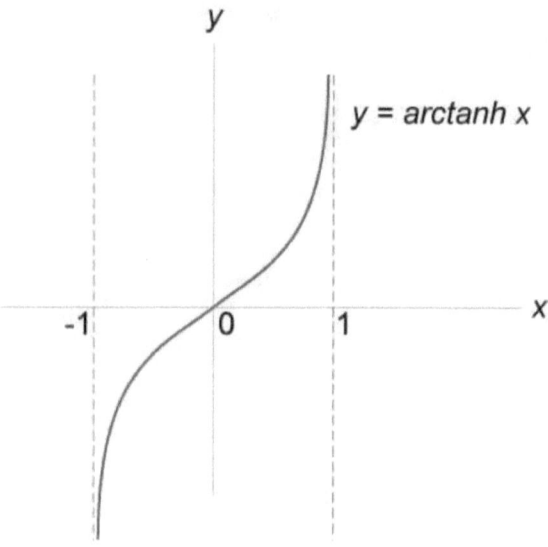

Figure 171.

746. Inverse Hyperbolic Cotangent Function
$y = \operatorname{arccoth} x$, $x \in (-\infty, -1) \cup (1, \infty)$.

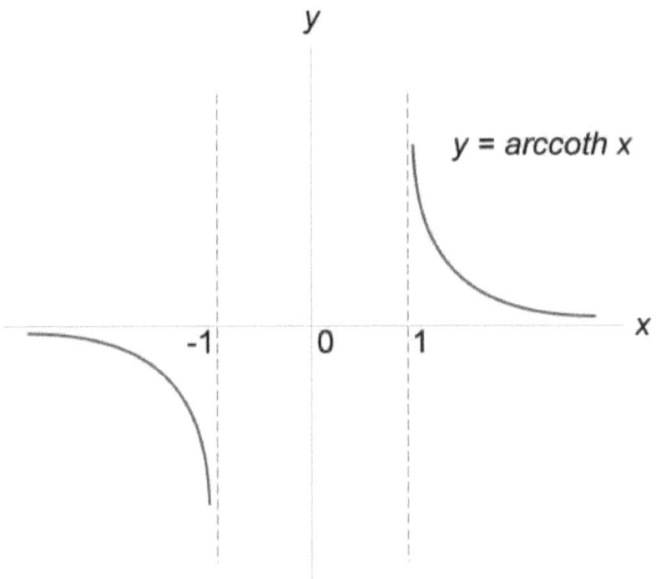

Figure 172.

747. Inverse Hyperbolic Secant Function
$y = \text{arcsech } x$, $x \in (0, 1]$.

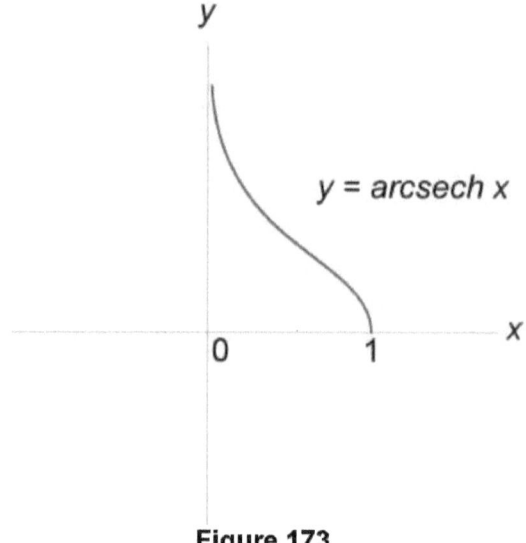

Figure 173.

748. Inverse Hyperbolic Cosecant Function
$y = \text{arccsch } x$, $x \in \mathbb{R}$, $x \neq 0$.

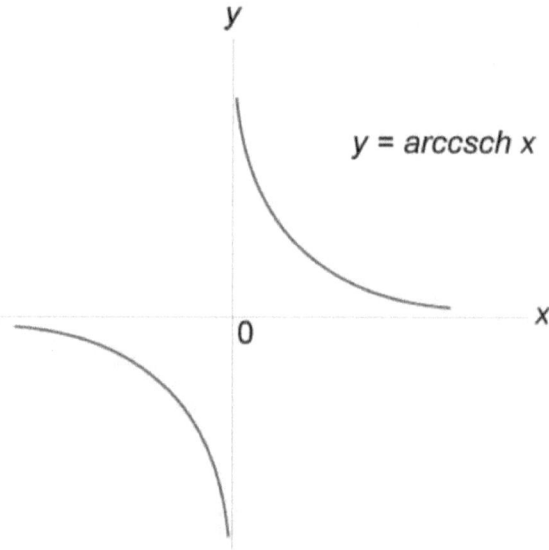

Figure 174.

CHAPTER 8. DIFFERENTIAL CALCULUS

8.2 Limits of Functions

Functions: f(x), g(x)
Argument: x
Real constants: a, k

749. $\lim\limits_{x \to a}[f(x)+g(x)] = \lim\limits_{x \to a} f(x) + \lim\limits_{x \to a} g(x)$

750. $\lim\limits_{x \to a}[f(x)-g(x)] = \lim\limits_{x \to a} f(x) - \lim\limits_{x \to a} g(x)$

751. $\lim\limits_{x \to a}[f(x) \cdot g(x)] = \lim\limits_{x \to a} f(x) \cdot \lim\limits_{x \to a} g(x)$

752. $\lim\limits_{x \to a} \dfrac{f(x)}{g(x)} = \dfrac{\lim\limits_{x \to a} f(x)}{\lim\limits_{x \to a} g(x)}$, if $\lim\limits_{x \to a} g(x) \neq 0$.

753. $\lim\limits_{x \to a}[kf(x)] = k \lim\limits_{x \to a} f(x)$

754. $\lim\limits_{x \to a} f(g(x)) = f\!\left(\lim\limits_{x \to a} g(x)\right)$

755. $\lim\limits_{x \to a} f(x) = f(a)$, if the function $f(x)$ is continuous at $x = a$.

756. $\lim\limits_{x \to 0} \dfrac{\sin x}{x} = 1$

757. $\lim\limits_{x \to 0} \dfrac{\tan x}{x} = 1$

758. $\lim\limits_{x \to 0} \dfrac{\sin^{-1} x}{x} = 1$

759. $\lim\limits_{x \to 0} \dfrac{\tan^{-1} x}{x} = 1$

760. $\lim\limits_{x \to 0} \dfrac{\ln(1+x)}{x} = 1$

761. $\lim\limits_{x \to \infty} \left(1 + \dfrac{1}{x}\right)^x = e$

762. $\lim\limits_{x \to \infty} \left(1 + \dfrac{k}{x}\right)^x = e^k$

763. $\lim\limits_{x \to 0} a^x = 1$

8.3 Definition and Properties of the Derivative

Functions: f, g, y, u, v
Independent variable: x
Real constant: k
Angle: α

764. $y'(x) = \lim\limits_{\Delta x \to 0} \dfrac{f(x + \Delta x) - f(x)}{\Delta x} = \lim\limits_{\Delta x \to 0} \dfrac{\Delta y}{\Delta x} = \dfrac{dy}{dx}$

CHAPTER 8. DIFFERENTIAL CALCULUS

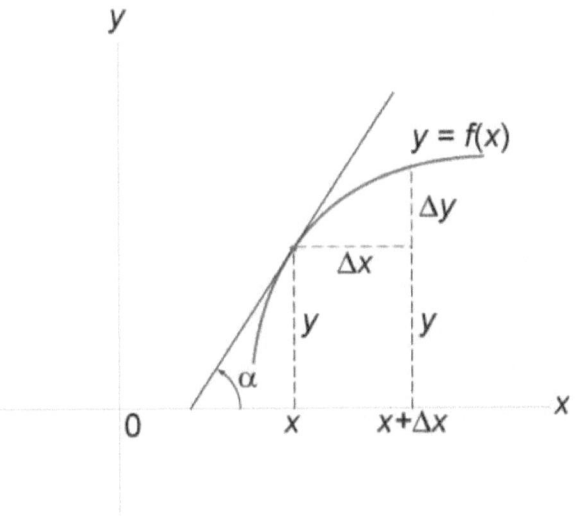

Figure 175.

765. $\dfrac{dy}{dx} = \tan \alpha$

766. $\dfrac{d(u+v)}{dx} = \dfrac{du}{dx} + \dfrac{dv}{dx}$

767. $\dfrac{d(u-v)}{dx} = \dfrac{du}{dx} - \dfrac{dv}{dx}$

768. $\dfrac{d(ku)}{dx} = k\dfrac{du}{dx}$

769. Product Rule
$\dfrac{d(u \cdot v)}{dx} = \dfrac{du}{dx} \cdot v + u \cdot \dfrac{dv}{dx}$

770. Quotient Rule

$$\frac{d}{dx}\left(\frac{u}{v}\right) = \frac{\frac{du}{dx}\cdot v - u\cdot \frac{dv}{dx}}{v^2}$$

771. Chain Rule

$y = f(g(x))$, $u = g(x)$,

$$\frac{dy}{dx} = \frac{dy}{du}\cdot\frac{du}{dx}.$$

772. Derivative of Inverse Function

$$\frac{dy}{dx} = \frac{1}{\frac{dx}{dy}},$$

where $x(y)$ is the inverse function of $y(x)$.

773. Reciprocal Rule

$$\frac{d}{dx}\left(\frac{1}{y}\right) = -\frac{\frac{dy}{dx}}{y^2}$$

774. Logarithmic Differentiation

$y = f(x)$, $\ln y = \ln f(x)$,

$$\frac{dy}{dx} = f(x)\cdot\frac{d}{dx}[\ln f(x)].$$

8.4 Table of Derivatives

Independent variable: x
Real constants: C, a, b, c
Natural number: n

CHAPTER 8. DIFFERENTIAL CALCULUS

775. $\dfrac{d}{dx}(C)=0$

776. $\dfrac{d}{dx}(x)=1$

777. $\dfrac{d}{dx}(ax+b)=a$

778. $\dfrac{d}{dx}(ax^2+bx+c)=ax+b$

779. $\dfrac{d}{dx}(x^n)=nx^{n-1}$

780. $\dfrac{d}{dx}(x^{-n})=-\dfrac{n}{x^{n+1}}$

781. $\dfrac{d}{dx}\left(\dfrac{1}{x}\right)=-\dfrac{1}{x^2}$

782. $\dfrac{d}{dx}(\sqrt{x})=\dfrac{1}{2\sqrt{x}}$

783. $\dfrac{d}{dx}(\sqrt[n]{x})=\dfrac{1}{n\sqrt[n]{x^{n-1}}}$

784. $\dfrac{d}{dx}(\ln x)=\dfrac{1}{x}$

785. $\dfrac{d}{dx}(\log_a x)=\dfrac{1}{x\ln a}$, $a>0$, $a\neq 1$.

CHAPTER 8. DIFFERENTIAL CALCULUS

786. $\dfrac{d}{dx}(a^x) = a^x \ln a$, $a > 0$, $a \neq 1$.

787. $\dfrac{d}{dx}(e^x) = e^x$

788. $\dfrac{d}{dx}(\sin x) = \cos x$

789. $\dfrac{d}{dx}(\cos x) = -\sin x$

790. $\dfrac{d}{dx}(\tan x) = \dfrac{1}{\cos^2 x} = \sec^2 x$

791. $\dfrac{d}{dx}(\cot x) = -\dfrac{1}{\sin^2 x} = -\csc^2 x$

792. $\dfrac{d}{dx}(\sec x) = \tan x \cdot \sec x$

793. $\dfrac{d}{dx}(\csc x) = -\cot x \cdot \csc x$

794. $\dfrac{d}{dx}(\arcsin x) = \dfrac{1}{\sqrt{1-x^2}}$

795. $\dfrac{d}{dx}(\arccos x) = -\dfrac{1}{\sqrt{1-x^2}}$

796. $\dfrac{d}{dx}(\arctan x) = \dfrac{1}{1+x^2}$

797. $\dfrac{d}{dx}(\operatorname{arccot} x) = -\dfrac{1}{1+x^2}$

798. $\dfrac{d}{dx}(\operatorname{arcsec} x) = \dfrac{1}{|x|\sqrt{x^2-1}}$

799. $\dfrac{d}{dx}(\operatorname{arccsc} x) = -\dfrac{1}{|x|\sqrt{x^2-1}}$

800. $\dfrac{d}{dx}(\sinh x) = \cosh x$

801. $\dfrac{d}{dx}(\cosh x) = \sinh x$

802. $\dfrac{d}{dx}(\tanh x) = \dfrac{1}{\cosh^2 x} = \operatorname{sech}^2 x$

803. $\dfrac{d}{dx}(\coth x) = -\dfrac{1}{\sinh^2 x} = -\operatorname{csch}^2 x$

804. $\dfrac{d}{dx}(\operatorname{sech} x) = -\operatorname{sech} x \cdot \tanh x$

805. $\dfrac{d}{dx}(\operatorname{csch} x) = -\operatorname{csch} x \cdot \coth x$

806. $\dfrac{d}{dx}(\operatorname{arcsinh} x) = \dfrac{1}{\sqrt{x^2+1}}$

807. $\dfrac{d}{dx}(\operatorname{arccosh} x) = \dfrac{1}{\sqrt{x^2-1}}$

808. $\dfrac{d}{dx}(\operatorname{arctanh} x) = \dfrac{1}{1-x^2}$, $|x| < 1$.

809. $\dfrac{d}{dx}(\operatorname{arccoth} x) = -\dfrac{1}{x^2-1}$, $|x| > 1$.

810. $\dfrac{d}{dx}(u^v) = vu^{v-1} \cdot \dfrac{du}{dx} + u^v \ln u \cdot \dfrac{dv}{dx}$

8.5 Higher Order Derivatives

Functions: f, y, u, v
Independent variable: x
Natural number: n

811. Second derivative
$$f'' = (f')' = \left(\dfrac{dy}{dx}\right)' = \dfrac{d}{dx}\left(\dfrac{dy}{dx}\right) = \dfrac{d^2 y}{dx^2}$$

812. Higher-Order derivative
$$f^{(n)} = \dfrac{d^n y}{dx^n} = y^{(n)} = \left(f^{(n-1)}\right)'$$

813. $(u+v)^{(n)} = u^{(n)} + v^{(n)}$

814. $(u-v)^{(n)} = u^{(n)} - v^{(n)}$

815. Leibnitz's Formulas
$$(uv)'' = u''v + 2u'v' + uv''$$

CHAPTER 8. DIFFERENTIAL CALCULUS

$$(uv)''' = u'''v + 3u''v' + 3u'v'' + uv'''$$
$$(uv)^{(n)} = u^{(n)}v + nu^{(n-1)}v' + \frac{n(n-1)}{1 \cdot 2}u^{(n-2)}v'' + \ldots + uv^{(n)}$$

816. $\left(x^m\right)^{(n)} = \dfrac{m!}{(m-n)!}x^{m-n}$

817. $\left(x^n\right)^{(n)} = n!$

818. $\left(\log_a x\right)^{(n)} = \dfrac{(-1)^{n-1}(n-1)!}{x^n \ln a}$

819. $\left(\ln x\right)^{(n)} = \dfrac{(-1)^{n-1}(n-1)!}{x^n}$

820. $\left(a^x\right)^{(n)} = a^x \ln^n a$

821. $\left(e^x\right)^{(n)} = e^x$

822. $\left(a^{mx}\right)^{(n)} = m^n a^{mx} \ln^n a$

823. $\left(\sin x\right)^{(n)} = \sin\left(x + \dfrac{n\pi}{2}\right)$

824. $\left(\cos x\right)^{(n)} = \cos\left(x + \dfrac{n\pi}{2}\right)$

8.6 Applications of Derivative

Functions: f, g, y
Position of an object: s
Velocity: v
Acceleration: w
Independent variable: x
Time: t
Natural number: n

825. Velocity and Acceleration
$s = f(t)$ is the position of an object relative to a fixed coordinate system at a time t,
$v = s' = f'(t)$ is the instantaneous velocity of the object,
$w = v' = s'' = f''(t)$ is the instantaneous acceleration of the object.

826. Tangent Line
$$y - y_0 = f'(x_0)(x - x_0)$$

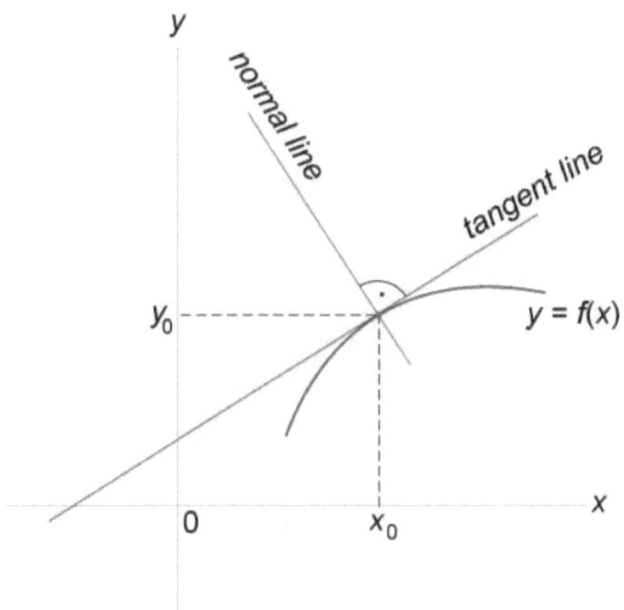

Figure 176.

827. Normal Line
$$y - y_0 = -\frac{1}{f'(x_0)}(x - x_0) \text{ (Fig 176)}$$

828. Increasing and Decreasing Functions.
If $f'(x_0) > 0$, then f(x) is increasing at x_0. (Fig 177, $x < x_1$, $x_2 < x$),
If $f'(x_0) < 0$, then f(x) is decreasing at x_0. (Fig 177, $x_1 < x < x_2$),
If $f'(x_0)$ does not exist or is zero, then the test fails.

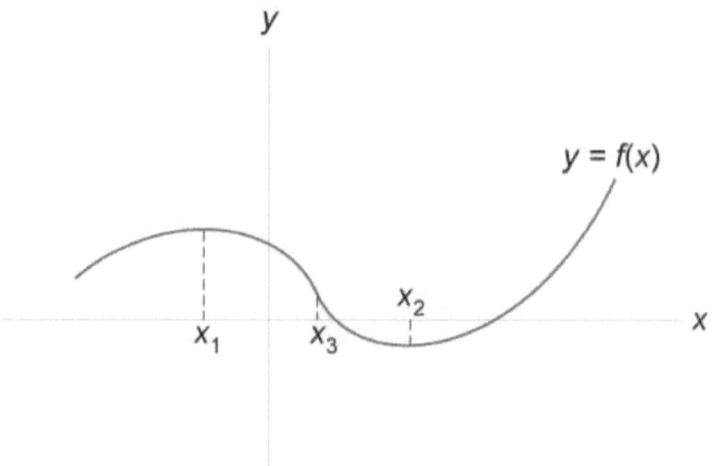

Figure 177.

829. Local extrema
A function f(x) has a local maximum at x_1 if and only if there exists some interval containing x_1 such that $f(x_1) \geq f(x)$ for all x in the interval (Fig.177).

A function f(x) has a local minimum at x_2 if and only if there exists some interval containing x_2 such that $f(x_2) \leq f(x)$ for all x in the interval (Fig.177).

830. Critical Points
A critical point on f(x) occurs at x_0 if and only if either $f'(x_0)$ is zero or the derivative doesn't exist.

831. First Derivative Test for Local Extrema.
If f(x) is increasing ($f'(x) > 0$) for all x in some interval $(a, x_1]$ and f(x) is decreasing ($f'(x) < 0$) for all x in some interval $[x_1, b)$, then f(x) has a local maximum at x_1 (Fig.177).

832. If f(x) is decreasing ($f'(x)<0$) for all x in some interval $(a, x_2]$ and f(x) is increasing ($f'(x)>0$) for all x in some interval $[x_2, b)$, then f(x) has a local minimum at x_2. (Fig.177).

833. Second Derivative Test for Local Extrema.
If $f'(x_1)=0$ and $f''(x_1)<0$, then f(x) has a local maximum at x_1.
If $f'(x_2)=0$ and $f''(x_2)>0$, then f(x) has a local minimum at x_2. (Fig.177)

834. Concavity.
f(x) is concave upward at x_0 if and only if $f'(x)$ is increasing at x_0 (Fig.177, $x_3 < x$).
f(x) is concave downward at x_0 if and only if $f'(x)$ is decreasing at x_0. (Fig.177, $x < x_3$).

835. Second Derivative Test for Concavity.
If $f''(x_0)>0$, then f(x) is concave upward at x_0.
If $f''(x_0)<0$, then f(x) is concave downward at x_0.
If $f''(x)$ does not exist or is zero, then the test fails.

836. Inflection Points
If $f'(x_3)$ exists and $f''(x)$ changes sign at $x=x_3$, then the point $(x_3, f(x_3))$ is an inflection point of the graph of f(x). If $f''(x_3)$ exists at the inflection point, then $f''(x_3)=0$ (Fig.177).

837. L'Hopital's Rule
$$\lim_{x \to c} \frac{f(x)}{g(x)} = \lim_{x \to c} \frac{f'(x)}{g'(x)} \text{ if } \lim_{x \to c} f(x) = \lim_{x \to c} g(x) = \begin{cases} 0 \\ \infty \end{cases}.$$

8.7 Differential

Functions: f, u, v
Independent variable: x
Derivative of a function: $y'(x)$, $f'(x)$
Real constant: C
Differential of function $y = f(x)$: dy
Differential of x: dx
Small change in x: Δx
Small change in y: Δy

838. $dy = y'dx$

839. $f(x + \Delta x) = f(x) + f'(x)\Delta x$

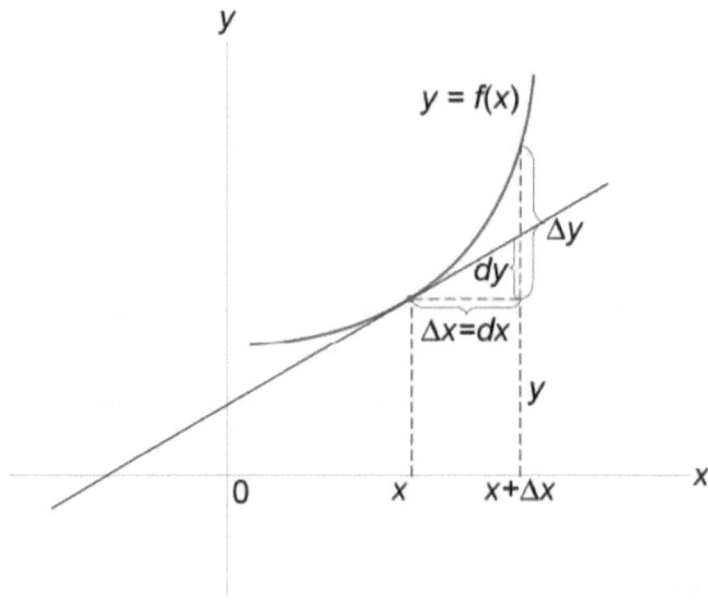

Figure 178.

840. Small Change in y
$$\Delta y = f(x + \Delta x) - f(x)$$

841. $d(u+v) = du + dv$

842. $d(u-v) = du - dv$

843. $d(Cu) = C\,du$

844. $d(uv) = v\,du + u\,dv$

845. $d\left(\dfrac{u}{v}\right) = \dfrac{v\,du - u\,dv}{v^2}$

8.8 Multivariable Functions

Functions of two variables: $z(x,y)$, $f(x,y)$, $g(x,y)$, $h(x,y)$
Arguments: x, y, t
Small changes in x, y, z, respectively: Δx, Δy, Δz.

846. First Order Partial Derivatives
The partial derivative with respect to x
$$\dfrac{\partial f}{\partial x} = f_x \quad \text{(also } \dfrac{\partial z}{\partial x} = z_x\text{)},$$
The partial derivative with respect to y
$$\dfrac{\partial f}{\partial y} = f_y \quad \text{(also } \dfrac{\partial z}{\partial y} = z_y\text{)}.$$

847. Second Order Partial Derivatives

$$\frac{\partial}{\partial x}\left(\frac{\partial f}{\partial x}\right) = \frac{\partial^2 f}{\partial x^2} = f_{xx},$$

$$\frac{\partial}{\partial y}\left(\frac{\partial f}{\partial y}\right) = \frac{\partial^2 f}{\partial y^2} = f_{yy},$$

$$\frac{\partial}{\partial y}\left(\frac{\partial f}{\partial x}\right) = \frac{\partial^2 f}{\partial y \partial x} = f_{xy},$$

$$\frac{\partial}{\partial x}\left(\frac{\partial f}{\partial y}\right) = \frac{\partial^2 f}{\partial x \partial y} = f_{yx}.$$

If the derivatives are continuous, then

$$\frac{\partial^2 f}{\partial y \partial x} = \frac{\partial^2 f}{\partial x \partial y}.$$

848. Chain Rules

If $f(x,y) = g(h(x,y))$ (g is a function of one variable h), then

$$\frac{\partial f}{\partial x} = g'(h(x,y))\frac{\partial h}{\partial x}, \quad \frac{\partial f}{\partial y} = g'(h(x,y))\frac{\partial h}{\partial y}.$$

If $h(t) = f(x(t), y(t))$, then $h'(t) = \frac{\partial f}{\partial x}\frac{dx}{dt} + \frac{\partial f}{\partial y}\frac{dy}{dt}.$

If $z = f(x(u,v), y(u,v))$, then

$$\frac{\partial z}{\partial u} = \frac{\partial f}{\partial x}\frac{\partial x}{\partial u} + \frac{\partial f}{\partial y}\frac{\partial y}{\partial u}, \quad \frac{\partial z}{\partial v} = \frac{\partial f}{\partial x}\frac{\partial x}{\partial v} + \frac{\partial f}{\partial y}\frac{\partial y}{\partial v}.$$

849. Small Changes

$$\Delta z \approx \frac{\partial f}{\partial x}\Delta x + \frac{\partial f}{\partial y}\Delta y$$

CHAPTER 8. DIFFERENTIAL CALCULUS

850. Local Maxima and Minima
$f(x,y)$ has a local maximum at (x_0,y_0) if $f(x,y) \leq f(x_0,y_0)$ for all (x,y) sufficiently close to (x_0,y_0).

$f(x,y)$ has a local minimum at (x_0,y_0) if $f(x,y) \geq f(x_0,y_0)$ for all (x,y) sufficiently close to (x_0,y_0).

851. Stationary Points
$$\frac{\partial f}{\partial x} = \frac{\partial f}{\partial y} = 0.$$
Local maxima and local minima occur at stationary points.

852. Saddle Point
A stationary point which is neither a local maximum nor a local minimum

853. Second Derivative Test for Stationary Points
Let (x_0,y_0) be a stationary point ($\frac{\partial f}{\partial x} = \frac{\partial f}{\partial y} = 0$).
$$D = \begin{vmatrix} f_{xx}(x_0,y_0) & f_{xy}(x_0,y_0) \\ f_{yx}(x_0,y_0) & f_{yy}(x_0,y_0) \end{vmatrix}.$$

If $D > 0$, $f_{xx}(x_0,y_0) > 0$, (x_0,y_0) is a point of local minima.
If $D > 0$, $f_{xx}(x_0,y_0) < 0$, (x_0,y_0) is a point of local maxima.
If $D < 0$, (x_0,y_0) is a saddle point.
If $D = 0$, the test fails.

854. Tangent Plane
The equation of the tangent plane to the surface $z = f(x,y)$ at (x_0,y_0,z_0) is
$$z - z_0 = f_x(x_0,y_0)(x - x_0) + f_y(x_0,y_0)(y - y_0).$$

855. Normal to Surface
The equation of the normal to the surface $z = f(x,y)$ at (x_0, y_0, z_0) is
$$\frac{x - x_0}{f_x(x_0, y_0)} = \frac{y - y_0}{f_y(x_0, y_0)} = \frac{z - z_0}{-1}.$$

8.9 Differential Operators

Unit vectors along the coordinate axes: $\vec{i}, \vec{j}, \vec{k}$
Scalar functions (scalar fields): $f(x,y,z)$, $u(x_1, x_2, \ldots, x_n)$
Gradient of a scalar field: grad u, ∇u
Directional derivative: $\dfrac{\partial f}{\partial l}$
Vector function (vector field): $\vec{F}(P, Q, R)$
Divergence of a vector field: div $\vec{F}$, $\nabla \cdot \vec{F}$
Curl of a vector field: curl $\vec{F}$, $\nabla \times \vec{F}$
Laplacian operator: ∇^2

856. Gradient of a Scalar Function
$$\text{grad } f = \nabla f = \left(\frac{\partial f}{\partial x}, \frac{\partial f}{\partial y}, \frac{\partial f}{\partial z} \right),$$

$$\text{grad } u = \nabla u = \left(\frac{\partial u}{\partial x_1}, \frac{\partial u}{\partial x_2}, \ldots, \frac{\partial u}{\partial x_n} \right).$$

857. Directional Derivative
$$\frac{\partial f}{\partial l} = \frac{\partial f}{\partial x} \cos \alpha + \frac{\partial f}{\partial y} \cos \beta + \frac{\partial f}{\partial z} \cos \gamma,$$

where the direction is defined by the vector
$\vec{l}(\cos\alpha, \cos\beta, \cos\gamma)$, $\cos^2\alpha + \cos^2\beta + \cos^2\gamma = 1$.

858. Divergence of a Vector Field
$$\operatorname{div} \vec{F} = \nabla \cdot \vec{F} = \frac{\partial P}{\partial x} + \frac{\partial Q}{\partial y} + \frac{\partial R}{\partial z}$$

859. Curl of a Vector Field
$$\operatorname{curl} \vec{F} = \nabla \times \vec{F} = \begin{vmatrix} \vec{i} & \vec{j} & \vec{k} \\ \dfrac{\partial}{\partial x} & \dfrac{\partial}{\partial x} & \dfrac{\partial}{\partial x} \\ P & Q & R \end{vmatrix}$$
$$= \left(\frac{\partial R}{\partial y} - \frac{\partial Q}{\partial z}\right)\vec{i} + \left(\frac{\partial P}{\partial z} - \frac{\partial R}{\partial x}\right)\vec{j} + \left(\frac{\partial Q}{\partial x} - \frac{\partial P}{\partial y}\right)\vec{k}$$

860. Laplacian Operator
$$\nabla^2 f = \frac{\partial^2 f}{\partial x^2} + \frac{\partial^2 f}{\partial y^2} + \frac{\partial^2 f}{\partial z^2}$$

861. $\operatorname{div}(\operatorname{curl} \vec{F}) = \nabla \cdot (\nabla \times \vec{F}) \equiv 0$

862. $\operatorname{curl}(\operatorname{grad} f) = \nabla \times (\nabla f) \equiv 0$

863. $\operatorname{div}(\operatorname{grad} f) = \nabla \cdot (\nabla f) = \nabla^2 f$

864. $\operatorname{curl}(\operatorname{curl} \vec{F}) = \operatorname{grad}(\operatorname{div} \vec{F}) - \nabla^2 \vec{F} = \nabla(\nabla \cdot \vec{F}) - \nabla^2 \vec{F}$

Chapter 9
Integral Calculus

Functions: f, g, u, v
Independent variables: x, t, ξ
Indefinite integral of a function: $\int f(x)dx$, $\int g(x)dx$, ...
Derivative of a function: y'(x), f'(x), F'(x), ...
Real constants: C, a, b, c, d, k
Natural numbers: m, n, i, j

9.1 Indefinite Integral

865. $\int f(x)dx = F(x) + C$ if $F'(x) = f(x)$.

866. $\left(\int f(x)dx\right)' = f(x)$

867. $\int kf(x)dx = k\int f(x)dx$

868. $\int [f(x)+g(x)]dx = \int f(x)dx + \int g(x)dx$

869. $\int [f(x)-g(x)]dx = \int f(x)dx - \int g(x)dx$

870. $\int f(ax)dx = \dfrac{1}{a}F(ax) + C$

CHAPTER 9. INTEGRAL CALCULUS

871. $\int f(ax+b)dx = \dfrac{1}{a}F(ax+b)+C$

872. $\int f(x)f'(x)dx = \dfrac{1}{2}f^2(x)+C$

873. $\int \dfrac{f'(x)}{f(x)}dx = \ln|f(x)|+C$

874. **Method of Substitution**
$\int f(x)dx = \int f(u(t))u'(t)dt$ if $x = u(t)$.

875. **Integration by Parts**
$\int u\,dv = uv - \int v\,du$,
where $u(x)$, $v(x)$ are differentiable functions.

9.2 Integrals of Rational Functions

876. $\int a\,dx = ax + C$

877. $\int x\,dx = \dfrac{x^2}{2}+C$

878. $\int x^2 dx = \dfrac{x^3}{3}+C$

879. $\int x^p dx = \dfrac{x^{p+1}}{p+1}+C$, $p \neq -1$.

CHAPTER 9. INTEGRAL CALCULUS

880. $\int (ax+b)^n \, dx = \dfrac{(ax+b)^{n+1}}{a(n+1)} + C, \; n \neq -1.$

881. $\int \dfrac{dx}{x} = \ln|x| + C$

882. $\int \dfrac{dx}{ax+b} = \dfrac{1}{a} \ln|ax+b| + C$

883. $\int \dfrac{ax+b}{cx+d} \, dx = \dfrac{a}{c} x + \dfrac{bc-ad}{c^2} \ln|cx+d| + C$

884. $\int \dfrac{dx}{(x+a)(x+b)} = \dfrac{1}{a-b} \ln\left|\dfrac{x+b}{x+a}\right| + C, \; a \neq b.$

885. $\int \dfrac{x \, dx}{a+bx} = \dfrac{1}{b^2}(a+bx - a\ln|a+bx|) + C$

886. $\int \dfrac{x^2 \, dx}{a+bx} = \dfrac{1}{b^3}\left[\dfrac{1}{2}(a+bx)^2 - 2a(a+bx) + a^2 \ln|a+bx|\right] + C$

887. $\int \dfrac{dx}{x(a+bx)} = \dfrac{1}{a} \ln\left|\dfrac{a+bx}{x}\right| + C$

888. $\int \dfrac{dx}{x^2(a+bx)} = -\dfrac{1}{ax} + \dfrac{b}{a^2} \ln\left|\dfrac{a+bx}{x}\right| + C$

889. $\int \dfrac{x \, dx}{(a+bx)^2} = \dfrac{1}{b^2}\left(\ln|a+bx| + \dfrac{a}{a+bx}\right) + C$

CHAPTER 9. INTEGRAL CALCULUS

890. $\int \dfrac{x^2 dx}{(a+bx)^2} = \dfrac{1}{b^3}\left(a+bx-2a\ln|a+bx|-\dfrac{a^2}{a+bx}\right)+C$

891. $\int \dfrac{dx}{x(a+bx)^2} = \dfrac{1}{a(a+bx)} + \dfrac{1}{a^2}\ln\left|\dfrac{a+bx}{x}\right|+C$

892. $\int \dfrac{dx}{x^2-1} = \dfrac{1}{2}\ln\left|\dfrac{x-1}{x+1}\right|+C$

893. $\int \dfrac{dx}{1-x^2} = \dfrac{1}{2}\ln\left|\dfrac{1+x}{1-x}\right|+C$

894. $\int \dfrac{dx}{a^2-x^2} = \dfrac{1}{2a}\ln\left|\dfrac{a+x}{a-x}\right|+C$

895. $\int \dfrac{dx}{x^2-a^2} = \dfrac{1}{2a}\ln\left|\dfrac{x-a}{x+a}\right|+C$

896. $\int \dfrac{dx}{1+x^2} = \tan^{-1}x+C$

897. $\int \dfrac{dx}{a^2+x^2} = \dfrac{1}{a}\tan^{-1}\dfrac{x}{a}+C$

898. $\int \dfrac{xdx}{x^2+a^2} = \dfrac{1}{2}\ln(x^2+a^2)+C$

899. $\int \dfrac{dx}{a+bx^2} = \dfrac{1}{\sqrt{ab}}\arctan\left(x\sqrt{\dfrac{b}{a}}\right)+C$, $ab>0$.

900. $\int \dfrac{x\,dx}{a+bx^2} = \dfrac{1}{2b}\ln\left|x^2+\dfrac{a}{b}\right|+C$

901. $\int \dfrac{dx}{x(a+bx^2)} = \dfrac{1}{2a}\ln\left|\dfrac{x^2}{a+bx^2}\right|+C$

902. $\int \dfrac{dx}{a^2-b^2x^2} = \dfrac{1}{2ab}\ln\left|\dfrac{a+bx}{a-bx}\right|+C$

903. $\int \dfrac{dx}{ax^2+bx+c} = \dfrac{1}{\sqrt{b^2-4ac}}\ln\left|\dfrac{2ax+b-\sqrt{b^2-4ac}}{2ax+b+\sqrt{b^2-4ac}}\right|+C$,

$b^2-4ac>0$.

904. $\int \dfrac{dx}{ax^2+bx+c} = \dfrac{2}{\sqrt{4ac-b^2}}\arctan\dfrac{2ax+b}{\sqrt{4ac-b^2}}+C$,

$b^2-4ac<0$.

9.3 Integrals of Irrational Functions

905. $\int \dfrac{dx}{\sqrt{ax+b}} = \dfrac{2}{a}\sqrt{ax+b}+C$

906. $\int \sqrt{ax+b}\,dx = \dfrac{2}{3a}(ax+b)^{3/2}+C$

907. $\int \dfrac{x\,dx}{\sqrt{ax+b}} = \dfrac{2(ax-2b)}{3a^2}\sqrt{ax+b}+C$

908. $\int x\sqrt{ax+b}\, dx = \dfrac{2(3ax-2b)}{15a^2}(ax+b)^{3/2} + C$

909. $\int \dfrac{dx}{(x+c)\sqrt{ax+b}} = \dfrac{1}{\sqrt{b-ac}} \ln\left|\dfrac{\sqrt{ax+b}-\sqrt{b-ac}}{\sqrt{ax+b}+\sqrt{b-ac}}\right| + C$,

$b-ac > 0$.

910. $\int \dfrac{dx}{(x+c)\sqrt{ax+b}} = \dfrac{1}{\sqrt{ac-b}} \arctan\sqrt{\dfrac{ax+b}{ac-b}} + C$,

$b-ac < 0$.

911. $\int \sqrt{\dfrac{ax+b}{cx+d}}\, dx = \dfrac{1}{c}\sqrt{(ax+b)(cx+d)} -$

$- \dfrac{ad-bc}{c\sqrt{ac}} \ln\left|\sqrt{a(cx+d)} + \sqrt{c(ax+b)}\right| + C$, $a > 0$.

912. $\int \sqrt{\dfrac{ax+b}{cx+d}}\, dx = \dfrac{1}{c}\sqrt{(ax+b)(cx+d)} -$

$- \dfrac{ad-bc}{c\sqrt{ac}} \arctan\sqrt{\dfrac{a(cx+d)}{c(ax+b)}} + C$, ($a<0$, $c>0$).

913. $\int x^2\sqrt{a+bx}\, dx = \dfrac{2(8a^2-12abx+15b^2x^2)}{105b^3}\sqrt{(a+bx)^3} + C$

914. $\int \dfrac{x^2\, dx}{\sqrt{a+bx}} = \dfrac{2(8a^2-4abx+3b^2x^2)}{15b^3}\sqrt{a+bx} + C$

915. $\int \dfrac{dx}{x\sqrt{a+bx}} = \dfrac{1}{\sqrt{a}} \ln\left|\dfrac{\sqrt{a+bx}-\sqrt{a}}{\sqrt{a+bx}+\sqrt{a}}\right| + C$, $a>0$.

916. $\int \dfrac{dx}{x\sqrt{a+bx}} = \dfrac{2}{\sqrt{-a}} \arctan \left| \dfrac{a+bx}{-a} \right| + C$, $a<0$.

917. $\int \sqrt{\dfrac{a-x}{b+x}}\, dx = \sqrt{(a-x)(b+x)} + (a+b)\arcsin\sqrt{\dfrac{x+b}{a+b}} + C$

918. $\int \sqrt{\dfrac{a+x}{b-x}}\, dx = -\sqrt{(a+x)(b-x)} - (a+b)\arcsin\sqrt{\dfrac{b-x}{a+b}} + C$

919. $\int \sqrt{\dfrac{1+x}{1-x}}\, dx = -\sqrt{1-x^2} + \arcsin x + C$

920. $\int \dfrac{dx}{\sqrt{(x-a)(b-a)}} = 2\arcsin\sqrt{\dfrac{x-a}{b-a}} + C$

921. $\int \sqrt{a+bx-cx^2}\, dx = \dfrac{2cx-b}{4c}\sqrt{a+bx-cx^2} +$
$+ \dfrac{b^2-4ac}{8\sqrt{c^3}} \arcsin \dfrac{2cx-b}{\sqrt{b^2+4ac}} + C$

922. $\int \dfrac{dx}{\sqrt{ax^2+bx+c}} = \dfrac{1}{\sqrt{a}} \ln\left|2ax+b+2\sqrt{a(ax^2+bx+c)}\right| + C$,
$a>0$.

923. $\int \dfrac{dx}{\sqrt{ax^2+bx+c}} = -\dfrac{1}{\sqrt{a}} \arcsin \dfrac{2ax+b}{\sqrt{b^2-4ac}} + C$, $a<0$.

924. $\int \sqrt{x^2+a^2}\, dx = \dfrac{x}{2}\sqrt{x^2+a^2} + \dfrac{a^2}{2}\ln\left|x+\sqrt{x^2+a^2}\right| + C$

CHAPTER 9. INTEGRAL CALCULUS

925. $\int x\sqrt{x^2+a^2}\,dx = \frac{1}{3}(x^2+a^2)^{3/2} + C$

926. $\int x^2\sqrt{x^2+a^2}\,dx = \frac{x}{8}(2x^2+a^2)\sqrt{x^2+a^2} -$
$\qquad - \frac{a^4}{8}\ln\left|x+\sqrt{x^2+a^2}\right| + C$

927. $\int \frac{\sqrt{x^2+a^2}}{x^2}\,dx = -\frac{\sqrt{x^2+a^2}}{x} + \ln\left|x+\sqrt{x^2+a^2}\right| + C$

928. $\int \frac{dx}{\sqrt{x^2+a^2}} = \ln\left|x+\sqrt{x^2+a^2}\right| + C$

929. $\int \frac{\sqrt{x^2+a^2}}{x}\,dx = \sqrt{x^2+a^2} + a\ln\left|\frac{x}{a+\sqrt{x^2+a^2}}\right| + C$

930. $\int \frac{x\,dx}{\sqrt{x^2+a^2}} = \sqrt{x^2+a^2} + C$

931. $\int \frac{x^2\,dx}{\sqrt{x^2+a^2}} = \frac{x}{2}\sqrt{x^2+a^2} - \frac{a^2}{2}\ln\left|x+\sqrt{x^2+a^2}\right| + C$

932. $\int \frac{dx}{x\sqrt{x^2+a^2}} = \frac{1}{a}\ln\left|\frac{x}{a+\sqrt{x^2+a^2}}\right| + C$

933. $\int \sqrt{x^2-a^2}\,dx = \frac{x}{2}\sqrt{x^2-a^2} - \frac{a^2}{2}\ln\left|x+\sqrt{x^2-a^2}\right| + C$

934. $\int x\sqrt{x^2-a^2}\,dx = \frac{1}{3}(x^2-a^2)^{3/2} + C$

CHAPTER 9. INTEGRAL CALCULUS

935. $\displaystyle\int\frac{\sqrt{x^2-a^2}}{x}dx = \sqrt{x^2-a^2} + a\arcsin\frac{a}{x} + C$

936. $\displaystyle\int\frac{\sqrt{x^2-a^2}}{x^2}dx = -\frac{\sqrt{x^2-a^2}}{x} + \ln\left|x+\sqrt{x^2-a^2}\right| + C$

937. $\displaystyle\int\frac{dx}{\sqrt{x^2-a^2}} = \ln\left|x+\sqrt{x^2-a^2}\right| + C$

938. $\displaystyle\int\frac{xdx}{\sqrt{x^2-a^2}} = \sqrt{x^2-a^2} + C$

939. $\displaystyle\int\frac{x^2dx}{\sqrt{x^2-a^2}} = \frac{x}{2}\sqrt{x^2-a^2} + \frac{a^2}{2}\ln\left|x+\sqrt{x^2-a^2}\right| + C$

940. $\displaystyle\int\frac{dx}{x\sqrt{x^2-a^2}} = -\frac{1}{a}\arcsin\frac{a}{x} + C$

941. $\displaystyle\int\frac{dx}{(x+a)\sqrt{x^2-a^2}} = \frac{1}{a}\sqrt{\frac{x-a}{x+a}} + C$

942. $\displaystyle\int\frac{dx}{(x-a)\sqrt{x^2-a^2}} = -\frac{1}{a}\sqrt{\frac{x+a}{x-a}} + C$

943. $\displaystyle\int\frac{dx}{x^2\sqrt{x^2-a^2}} = \frac{\sqrt{x^2-a^2}}{a^2 x} + C$

944. $\displaystyle\int\frac{dx}{(x^2-a^2)^{3/2}} = -\frac{x}{a^2\sqrt{x^2-a^2}} + C$

945. $\int (x^2-a^2)^{3/2} dx = -\dfrac{x}{8}(2x^2-5a^2)\sqrt{x^2-a^2} +$

$+\dfrac{3a^4}{8}\ln\left|x+\sqrt{x^2-a^2}\right|+C$

946. $\int \sqrt{a^2-x^2}\, dx = \dfrac{x}{2}\sqrt{a^2-x^2}+\dfrac{a^2}{2}\arcsin\dfrac{x}{a}+C$

947. $\int x\sqrt{a^2-x^2}\, dx = -\dfrac{1}{3}(a^2-x^2)^{3/2}+C$

948. $\int x^2\sqrt{a^2-x^2}\, dx = \dfrac{x}{8}(2x^2-a^2)\sqrt{a^2-x^2}+\dfrac{a^4}{8}\arcsin\dfrac{x}{a}+C$

949. $\int \dfrac{\sqrt{a^2-x^2}}{x}\, dx = \sqrt{a^2-x^2}+a\ln\left|\dfrac{x}{a+\sqrt{a^2-x^2}}\right|+C$

950. $\int \dfrac{\sqrt{a^2-x^2}}{x^2}\, dx = -\dfrac{\sqrt{a^2-x^2}}{x}-\arcsin\dfrac{x}{a}+C$

951. $\int \dfrac{dx}{\sqrt{1-x^2}} = \arcsin x + C$

952. $\int \dfrac{dx}{\sqrt{a^2-x^2}} = \sin\dfrac{x}{a}+C$

953. $\int \dfrac{x\,dx}{\sqrt{a^2-x^2}} = -\sqrt{a^2-x^2}+C$

954. $\int \dfrac{x^2\,dx}{\sqrt{a^2-x^2}} = -\dfrac{x}{2}\sqrt{a^2-x^2}+\dfrac{a^2}{2}\arcsin\dfrac{x}{a}+C$

955. $\displaystyle\int\frac{dx}{(x+a)\sqrt{a^2-x^2}}=-\frac{1}{2}\sqrt{\frac{a-x}{a+x}}+C$

956. $\displaystyle\int\frac{dx}{(x-a)\sqrt{a^2-x^2}}=-\frac{1}{2}\sqrt{\frac{a+x}{a-x}}+C$

957. $\displaystyle\int\frac{dx}{(x+b)\sqrt{a^2-x^2}}=\frac{1}{\sqrt{b^2-a^2}}\arcsin\frac{bx+a^2}{a(x+b)}+C,\ b>a$.

958. $\displaystyle\int\frac{dx}{(x+b)\sqrt{a^2-x^2}}=\frac{1}{\sqrt{a^2-b^2}}\ln\left|\frac{x+b}{\sqrt{a^2-b^2}\sqrt{a^2-x^2}+a^2+bx}\right|+C$,

$b<a$.

959. $\displaystyle\int\frac{dx}{x^2\sqrt{a^2-x^2}}=-\frac{\sqrt{a^2-x^2}}{a^2 x}+C$

960. $\displaystyle\int(a^2-x^2)^{3/2}dx=\frac{x}{8}(5a^2-2x^2)\sqrt{a^2-x^2}+\frac{3a^4}{8}\arcsin\frac{x}{a}+C$

961. $\displaystyle\int\frac{dx}{(a^2-x^2)^{3/2}}=\frac{x}{a^2\sqrt{a^2-x^2}}+C$

9.4 Integrals of Trigonometric Functions

962. $\displaystyle\int\sin x\,dx=-\cos x+C$

963. $\displaystyle\int\cos x\,dx=\sin x+C$

964. $\int \sin^2 x \, dx = \dfrac{x}{2} - \dfrac{1}{4}\sin 2x + C$

965. $\int \cos^2 x \, dx = \dfrac{x}{2} + \dfrac{1}{4}\sin 2x + C$

966. $\int \sin^3 x \, dx = \dfrac{1}{3}\cos^3 x - \cos x + C = \dfrac{1}{12}\cos 3x - \dfrac{3}{4}\cos x + C$

967. $\int \cos^3 x \, dx = \sin x - \dfrac{1}{3}\sin^3 x + C = \dfrac{1}{12}\sin 3x + \dfrac{3}{4}\sin x + C$

968. $\int \dfrac{dx}{\sin x} = \int \csc x \, dx = \ln\left|\tan\dfrac{x}{2}\right| + C$

969. $\int \dfrac{dx}{\cos x} = \int \sec x \, dx = \ln\left|\tan\left(\dfrac{x}{2} + \dfrac{\pi}{4}\right)\right| + C$

970. $\int \dfrac{dx}{\sin^2 x} = \int \csc^2 x \, dx = -\cot x + C$

971. $\int \dfrac{dx}{\cos^2 x} = \int \sec^2 x \, dx = \tan x + C$

972. $\int \dfrac{dx}{\sin^3 x} = \int \csc^3 x \, dx = -\dfrac{\cos x}{2\sin^2 x} + \dfrac{1}{2}\ln\left|\tan\dfrac{x}{2}\right| + C$

973. $\int \dfrac{dx}{\cos^3 x} = \int \sec^3 x \, dx = \dfrac{\sin x}{2\cos^2 x} + \dfrac{1}{2}\ln\left|\tan\left(\dfrac{x}{2} + \dfrac{\pi}{4}\right)\right| + C$

974. $\int \sin x \cos x \, dx = -\dfrac{1}{4}\cos 2x + C$

CHAPTER 9. INTEGRAL CALCULUS

975. $\int \sin^2 x \cos x \, dx = \dfrac{1}{3}\sin^3 x + C$

976. $\int \sin x \cos^2 x \, dx = -\dfrac{1}{3}\cos^3 x + C$

977. $\int \sin^2 x \cos^2 x \, dx = \dfrac{x}{8} - \dfrac{1}{32}\sin 4x + C$

978. $\int \tan x \, dx = -\ln|\cos x| + C$

979. $\int \dfrac{\sin x}{\cos^2 x} dx = \dfrac{1}{\cos x} + C = \sec x + C$

980. $\int \dfrac{\sin^2 x}{\cos x} dx = \ln\left|\tan\left(\dfrac{x}{2} + \dfrac{\pi}{4}\right)\right| - \sin x + C$

981. $\int \tan^2 x \, dx = \tan x - x + C$

982. $\int \cot x \, dx = \ln|\sin x| + C$

983. $\int \dfrac{\cos x}{\sin^2 x} dx = -\dfrac{1}{\sin x} + C = -\csc x + C$

984. $\int \dfrac{\cos^2 x}{\sin x} dx = \ln\left|\tan\dfrac{x}{2}\right| + \cos x + C$

985. $\int \cot^2 x \, dx = -\cot x - x + C$

986. $\int \dfrac{dx}{\cos x \sin x} = \ln|\tan x| + C$

987. $\int \dfrac{dx}{\sin^2 x \cos x} = -\dfrac{1}{\sin x} + \ln\left|\tan\left(\dfrac{x}{2} + \dfrac{\pi}{4}\right)\right| + C$

988. $\int \dfrac{dx}{\sin x \cos^2 x} = \dfrac{1}{\cos x} + \ln\left|\tan\dfrac{x}{2}\right| + C$

989. $\int \dfrac{dx}{\sin^2 x \cos^2 x} = \tan x - \cot x + C$

990. $\int \sin mx \sin nx\, dx = -\dfrac{\sin(m+n)x}{2(m+n)} + \dfrac{\sin(m-n)x}{2(m-n)} + C$,

$m^2 \neq n^2$.

991. $\int \sin mx \cos nx\, dx = -\dfrac{\cos(m+n)x}{2(m+n)} - \dfrac{\cos(m-n)x}{2(m-n)} + C$,

$m^2 \neq n^2$.

992. $\int \cos mx \cos nx\, dx = \dfrac{\sin(m+n)x}{2(m+n)} + \dfrac{\sin(m-n)x}{2(m-n)} + C$,

$m^2 \neq n^2$.

993. $\int \sec x \tan x\, dx = \sec x + C$

994. $\int \csc x \cot x\, dx = -\csc x + C$

995. $\int \sin x \cos^n x\, dx = -\dfrac{\cos^{n+1} x}{n+1} + C$

996. $\int \sin^n x \cos x\, dx = \dfrac{\sin^{n+1} x}{n+1} + C$

997. $\int \arcsin x \, dx = x \arcsin x + \sqrt{1-x^2} + C$

998. $\int \arccos x \, dx = x \arccos x - \sqrt{1-x^2} + C$

999. $\int \arctan x \, dx = x \arctan x - \frac{1}{2}\ln(x^2+1) + C$

1000. $\int \text{arc cot } x \, dx = x \text{ arc cot } x + \frac{1}{2}\ln(x^2+1) + C$

9.5 Integrals of Hyperbolic Functions

1001. $\int \sinh x \, dx = \cosh x + C$

1002. $\int \cosh x \, dx = \sinh x + C$

1003. $\int \tanh x \, dx = \ln \cosh x + C$

1004. $\int \coth x \, dx = \ln|\sinh x| + C$

1005. $\int \text{sech}^2 x \, dx = \tanh x + C$

1006. $\int \text{csch}^2 x \, dx = -\coth x + C$

1007. $\int \text{sech} x \tanh x \, dx = -\text{sech} x + C$

1008. $\int \operatorname{csch} x \coth x \, dx = -\operatorname{csch} x + C$

9.6 Integrals of Exponential and Logarithmic Functions

1009. $\int e^x \, dx = e^x + C$

1010. $\int a^x \, dx = \dfrac{a^x}{\ln a} + C$

1011. $\int e^{ax} \, dx = \dfrac{e^{ax}}{a} + C$

1012. $\int x e^{ax} \, dx = \dfrac{e^{ax}}{a^2}(ax - 1) + C$

1013. $\int \ln x \, dx = x \ln x - x + C$

1014. $\int \dfrac{dx}{x \ln x} = \ln|\ln x| + C$

1015. $\int x^n \ln x \, dx = x^{n+1} \left[\dfrac{\ln x}{n+1} - \dfrac{1}{(n+1)^2} \right] + C$

1016. $\int e^{ax} \sin bx \, dx = \dfrac{a \sin bx - b \cos bx}{a^2 + b^2} e^{ax} + C$

1017. $\int e^{ax} \cos bx \, dx = \dfrac{a \cos bx + b \sin bx}{a^2 + b^2} e^{ax} + C$

9.7 Reduction Formulas

1018. $\int x^n e^{mx} \, dx = \dfrac{1}{m} x^n e^{mx} - \dfrac{n}{m} \int x^{n-1} e^{mx} \, dx$

1019. $\int \dfrac{e^{mx}}{x^n} \, dx = -\dfrac{e^{mx}}{(n-1)x^{n-1}} + \dfrac{m}{n-1} \int \dfrac{e^{mx}}{x^{n-1}} \, dx, \ n \neq 1$.

1020. $\int \sinh^n x \, dx = \dfrac{1}{n} \sinh^{n-1} x \cosh x - \dfrac{n-1}{n} \int \sinh^{n-2} x \, dx$

1021. $\int \dfrac{dx}{\sinh^n x} = -\dfrac{\cosh x}{(n-1)\sinh^{n-1} x} - \dfrac{n-2}{n-1} \int \dfrac{dx}{\sinh^{n-2} x}, \ n \neq 1$.

1022. $\int \cosh^n x \, dx = \dfrac{1}{n} \sinh x \cosh^{n-1} x \cosh x + \dfrac{n-1}{n} \int \cosh^{n-2} x \, dx$

1023. $\int \dfrac{dx}{\cosh^n x} = -\dfrac{\sinh x}{(n-1)\cosh^{n-1} x} + \dfrac{n-2}{n-1} \int \dfrac{dx}{\cosh^{n-2} x}, \ n \neq 1$.

1024. $\int \sinh^n x \cosh^m x \, dx = \dfrac{\sinh^{n+1} x \cosh^{m-1} x}{n+m}$
$\qquad + \dfrac{m-1}{n+m} \int \sinh^n x \cosh^{m-2} x \, dx$

1025. $\int \sinh^n x \cosh^m x \, dx = \dfrac{\sinh^{n-1} x \cosh^{m+1} x}{n+m}$

$$-\frac{n-1}{n+m}\int\sinh^{n-2}x\cosh^m x\,dx$$

1026. $\int\tanh^n x\,dx = -\frac{1}{n-1}\tanh^{n-1}x + \int\tanh^{n-2}x\,dx$, $n\neq 1$.

1027. $\int\coth^n x\,dx = -\frac{1}{n-1}\coth^{n-1}x + \int\coth^{n-2}x\,dx$, $n\neq 1$.

1028. $\int\operatorname{sech}^n x\,dx = \frac{\operatorname{sech}^{n-2}x\tanh x}{n-1} + \frac{n-2}{n-1}\int\operatorname{sech}^{n-2}x\,dx$, $n\neq 1$.

1029. $\int\sin^n x\,dx = -\frac{1}{n}\sin^{n-1}x\cos x + \frac{n-1}{n}\int\sin^{n-2}x\,dx$

1030. $\int\frac{dx}{\sin^n x} = -\frac{\cos x}{(n-1)\sin^{n-1}x} + \frac{n-2}{n-1}\int\frac{dx}{\sin^{n-2}x}$, $n\neq 1$.

1031. $\int\cos^n x\,dx = \frac{1}{n}\sin x\cos^{n-1}x + \frac{n-1}{n}\int\cos^{n-2}x\,dx$

1032. $\int\frac{dx}{\cos^n x} = \frac{\sin x}{(n-1)\cos^{n-1}x} + \frac{n-2}{n-1}\int\frac{dx}{\cos^{n-2}x}$, $n\neq 1$.

1033. $\int\sin^n x\cos^m x\,dx = \frac{\sin^{n+1}x\cos^{m-1}x}{n+m}$

$$+\frac{m-1}{n+m}\int\sin^n x\cos^{m-2}x\,dx$$

1034. $\int\sin^n x\cos^m x\,dx = -\frac{\sin^{n-1}x\cos^{m+1}x}{n+m}$

$$+\frac{n-1}{n+m}\int\sin^{n-2}x\cos^m x\,dx$$

1035. $\int\tan^n x\,dx = \dfrac{1}{n-1}\tan^{n-1}x - \int\tan^{n-2}x\,dx$, $n\neq 1$.

1036. $\int\cot^n x\,dx = -\dfrac{1}{n-1}\cot^{n-1}x - \int\cot^{n-2}x\,dx$, $n\neq 1$.

1037. $\int\sec^n x\,dx = \dfrac{\sec^{n-2}x\tan x}{n-1} + \dfrac{n-2}{n-1}\int\sec^{n-2}x\,dx$, $n\neq 1$.

1038. $\int\csc^n x\,dx = -\dfrac{\csc^{n-2}x\cot x}{n-1} + \dfrac{n-2}{n-1}\int\csc^{n-2}x\,dx$, $n\neq 1$.

1039. $\int x^n \ln^m x\,dx = \dfrac{x^{n+1}\ln^m x}{n+1} - \dfrac{m}{n+1}\int x^n \ln^{m-1}x\,dx$

1040. $\int\dfrac{\ln^m x}{x^n}dx = -\dfrac{\ln^m x}{(n-1)x^{n-1}} + \dfrac{m}{n-1}\int\dfrac{\ln^{m-1}x}{x^n}dx$, $n\neq 1$.

1041. $\int\ln^n x\,dx = x\ln^n x - n\int\ln^{n-1}x\,dx$

1042. $\int x^n \sinh x\,dx = x^n\cosh x - n\int x^{n-1}\cosh x\,dx$

1043. $\int x^n \cosh x\,dx = x^n\sinh x - n\int x^{n-1}\sinh x\,dx$

1044. $\int x^n \sin x\,dx = -x^n\cos x + n\int x^{n-1}\cos x\,dx$

1045. $\int x^n \cos x\,dx = x^n\sin x - n\int x^{n-1}\sin x\,dx$

1046. $\int x^n \sin^{-1} x \, dx = \frac{x^{n+1}}{n+1} \sin^{-1} x - \frac{1}{n+1} \int \frac{x^{n+1}}{\sqrt{1-x^2}} dx$

1047. $\int x^n \cos^{-1} x \, dx = \frac{x^{n+1}}{n+1} \cos^{-1} x + \frac{1}{n+1} \int \frac{x^{n+1}}{\sqrt{1-x^2}} dx$

1048. $\int x^n \tan^{-1} x \, dx = \frac{x^{n+1}}{n+1} \tan^{-1} x - \frac{1}{n+1} \int \frac{x^{n+1}}{1+x^2} dx$

1049. $\int \frac{x^n dx}{ax^n + b} = \frac{x}{a} - \frac{b}{a} \int \frac{dx}{ax^n + b}$

1050. $\int \frac{dx}{(ax^2 + bx + c)^n} = \frac{-2ax - b}{(n-1)(b^2 - 4ac)(ax^2 + bx + c)^{n-1}}$
$- \frac{2(2n-3)a}{(n-1)(b^2 - 4ac)} \int \frac{dx}{(ax^2 + bx + c)^{n-1}}, \; n \neq 1.$

1051. $\int \frac{dx}{(x^2 + a^2)^n} = \frac{x}{2(n-1)a^2(x^2 + a^2)^{n-1}} + \frac{2n-3}{2(n-1)a^2} \int \frac{dx}{(x^2 + a^2)^{n-1}},$
$n \neq 1.$

1052. $\int \frac{dx}{(x^2 - a^2)^n} = -\frac{x}{2(n-1)a^2(x^2 - a^2)^{n-1}}$
$- \frac{2n-3}{2(n-1)a^2} \int \frac{dx}{(x^2 - a^2)^{n-1}}, \; n \neq 1.$

9.8 Definite Integral

Definite integral of a function: $\int_a^b f(x)dx$, $\int_a^b g(x)dx$, ...

Riemann sum: $\sum_{i=1}^{n} f(\xi_i)\Delta x_i$

Small changes: Δx_i
Antiderivatives: $F(x)$, $G(x)$
Limits of integrations: a, b, c, d

1053. $\int_a^b f(x)dx = \lim_{\substack{n \to \infty \\ \max \Delta x_i \to 0}} \sum_{i=1}^{n} f(\xi_i)\Delta x_i$,

where $\Delta x_i = x_i - x_{i-1}$, $x_{i-1} \leq \xi_i \leq x_i$.

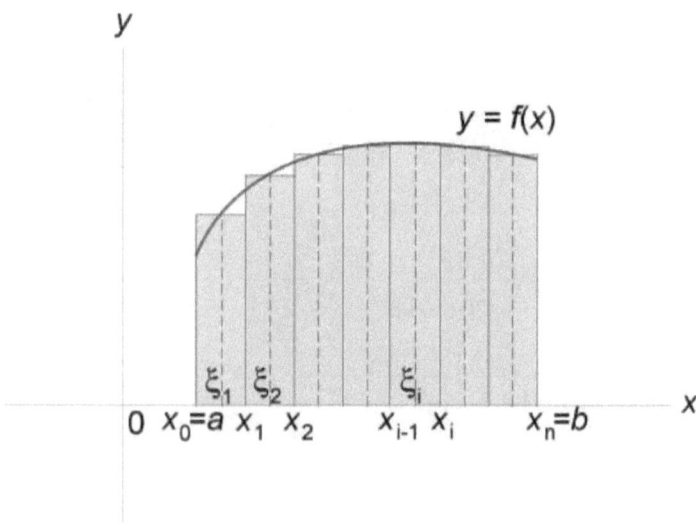

Figure 179.

CHAPTER 9. INTEGRAL CALCULUS

1054. $\int_a^b 1\,dx = b - a$

1055. $\int_a^b kf(x)\,dx = k\int_a^b f(x)\,dx$

1056. $\int_a^b [f(x) + g(x)]\,dx = \int_a^b f(x)\,dx + \int_a^b g(x)\,dx$

1057. $\int_a^b [f(x) - g(x)]\,dx = \int_a^b f(x)\,dx - \int_a^b g(x)\,dx$

1058. $\int_a^a f(x)\,dx = 0$

1059. $\int_a^b f(x)\,dx = -\int_b^a f(x)\,dx$

1060. $\int_a^b f(x)\,dx = \int_a^c f(x)\,dx + \int_c^b f(x)\,dx$ for $a < c < b$.

1061. $\int_a^b f(x)\,dx \geq 0$ if $f(x) \geq 0$ on $[a,b]$.

1062. $\int_a^b f(x)\,dx \leq 0$ if $f(x) \leq 0$ on $[a,b]$.

1063. Fundamental Theorem of Calculus

$$\int_a^b f(x)dx = F(x)\Big|_a^b = F(b)-F(a) \text{ if } F'(x)=f(x).$$

1064. Method of Substitution
If $x = g(t)$, then
$$\int_a^b f(x)dx = \int_c^d f(g(t))g'(t)dt,$$
where
$c = g^{-1}(a)$, $d = g^{-1}(b)$.

1065. Integration by Parts
$$\int_a^b udv = (uv)\Big|_a^b - \int_a^b vdu$$

1066. Trapezoidal Rule
$$\int_a^b f(x)dx = \frac{b-a}{2n}\left[f(x_0)+f(x_n)+2\sum_{i=1}^{n-1}f(x_i)\right]$$

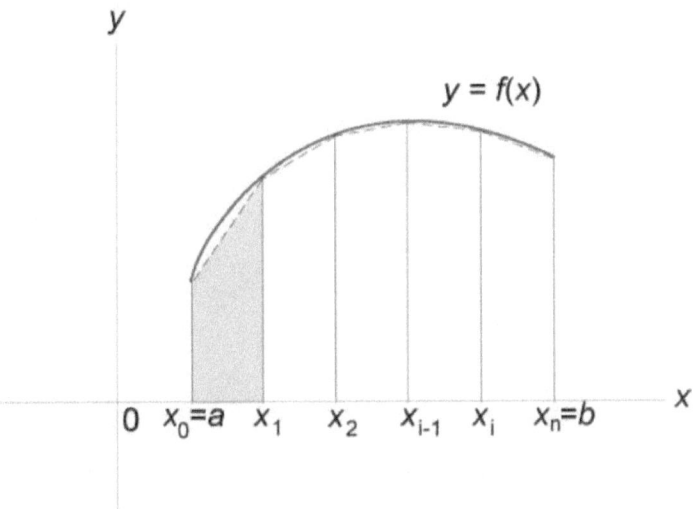

Figure 180.

1067. Simpson's Rule

$$\int_a^b f(x)dx = \frac{b-a}{3n}[f(x_0)+4f(x_1)+2f(x_2)+4f(x_3)+ \\ +2f(x_4)+\ldots+4f(x_{n-1})+f(x_n)],$$

where

$$x_i = a + \frac{b-a}{n}i, \ i = 0, 1, 2, \ldots, n.$$

CHAPTER 9. INTEGRAL CALCULUS

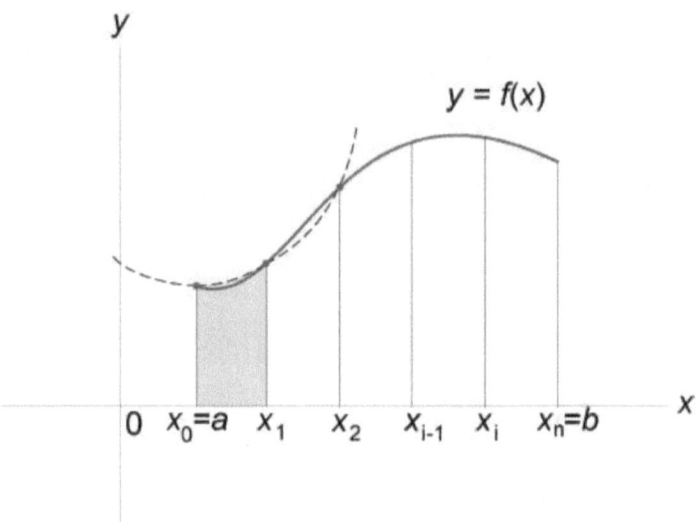

Figure 181.

1068. Area Under a Curve

$$S = \int_a^b f(x)dx = F(b) - F(a),$$

where $F'(x) = f(x)$.

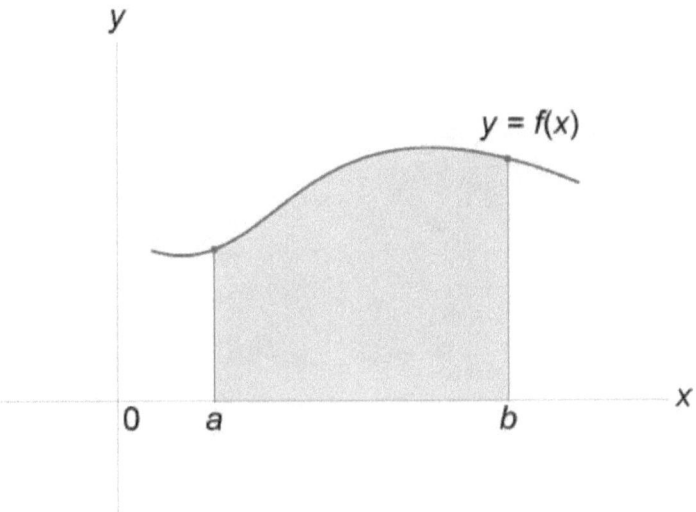

Figure 182.

1069. Area Between Two Curves
$$S = \int_a^b [f(x) - g(x)]dx = F(b) - G(b) - F(a) + G(a),$$
where $F'(x) = f(x)$, $G'(x) = g(x)$.

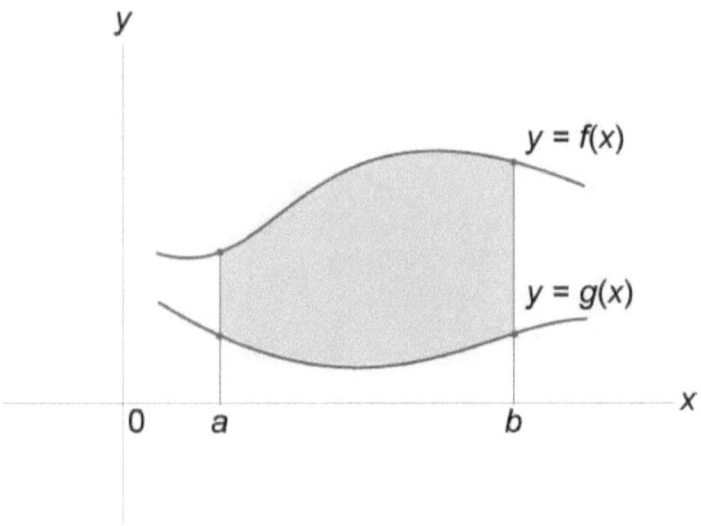

Figure 183.

9.9 Improper Integral

1070. The definite integral $\int_a^b f(x)dx$ is called an improper integral if
- a or b is infinite,
- $f(x)$ has one or more points of discontinuity in the interval $[a,b]$.

1071. If $f(x)$ is a continuous function on $[a,\infty)$, then
$$\int_a^\infty f(x)dx = \lim_{n\to\infty} \int_a^n f(x)dx.$$

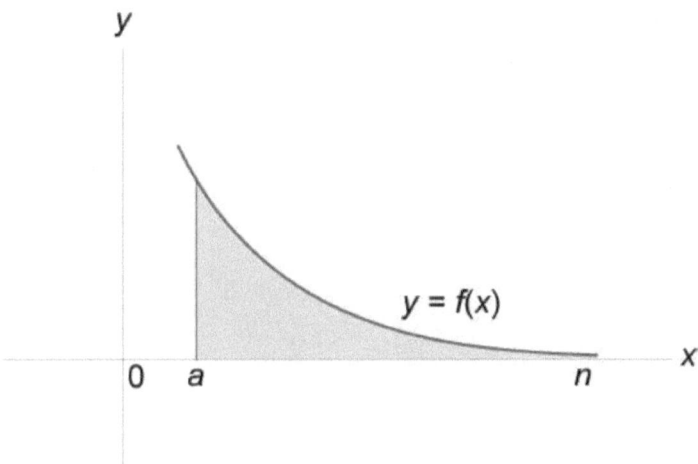

Figure 184.

1072. If $f(x)$ is a continuous function on $(-\infty, b]$, then
$$\int_{-\infty}^{b} f(x)dx = \lim_{n \to -\infty} \int_{n}^{b} f(x)dx.$$

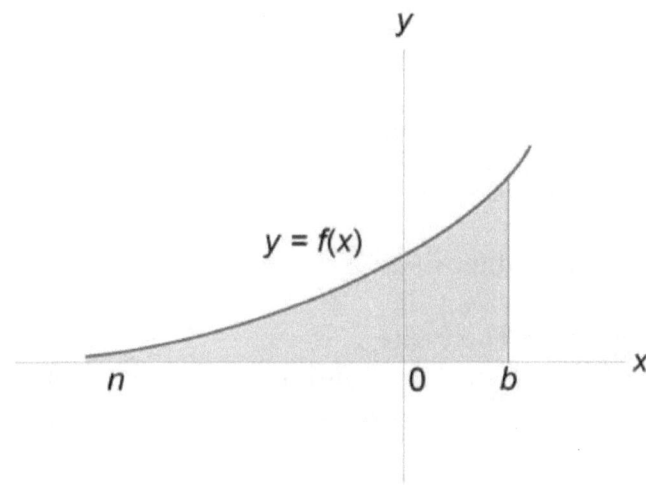

Figure 185.

Note : The improper integrals in 1071, 1072 are convergent if the limits exist and are finite; otherwise the integrals are divergent.

1073. $\int_{-\infty}^{\infty} f(x)dx = \int_{-\infty}^{c} f(x)dx + \int_{c}^{\infty} f(x)dx$

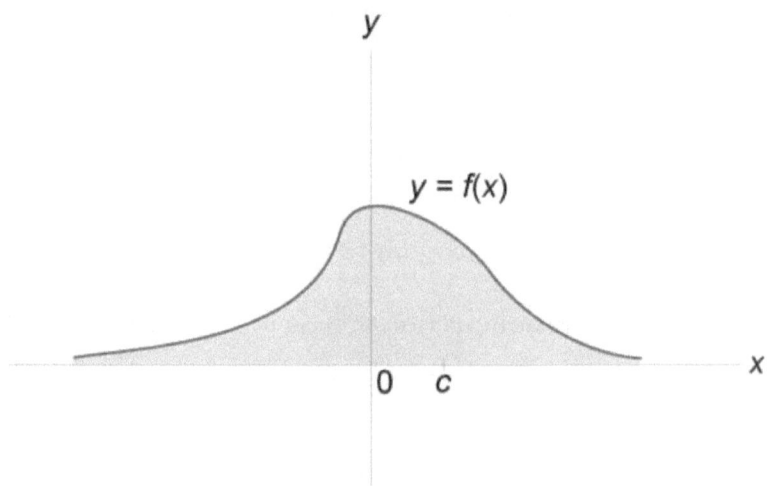

Figure 186.

If for some real number c, both of the integrals in the right side are convergent, then the integral $\int_{-\infty}^{\infty} f(x)dx$ is also convergent; otherwise it is divergent.

1074. **Comparison Theorems**

Let $f(x)$ and $g(x)$ be continuous functions on the closed interval $[a, \infty)$. Suppose that $0 \leq g(x) \leq f(x)$ for all x in $[a, \infty)$.

- If $\int_a^\infty f(x)dx$ is convergent, then $\int_a^\infty g(x)dx$ is also convergent,
- If $\int_a^\infty g(x)dx$ is divergent, then $\int_a^\infty f(x)dx$ is also divergent.

1075. Absolute Convergence

If $\int_a^\infty |f(x)|dx$ is convergent, then the integral $\int_a^\infty f(x)dx$ is absolutely convergent.

1076. Discontinuous Integrand

Let $f(x)$ be a function which is continuous on the interval $[a,b)$ but is discontinuous at $x=b$. Then

$$\int_a^b f(x)dx = \lim_{\varepsilon \to 0+} \int_a^{b-\varepsilon} f(x)dx$$

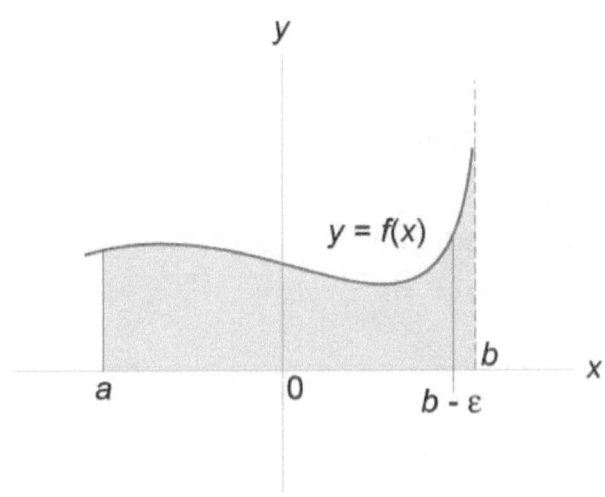

Figure 187.

1077. Let $f(x)$ be a continuous function for all real numbers x in the interval $[a,b]$ except for some point c in (a,b). Then

$$\int_a^b f(x)dx = \lim_{\varepsilon \to 0+} \int_a^{c-\varepsilon} f(x)dx + \lim_{\delta \to 0+} \int_{c+\delta}^b f(x)dx.$$

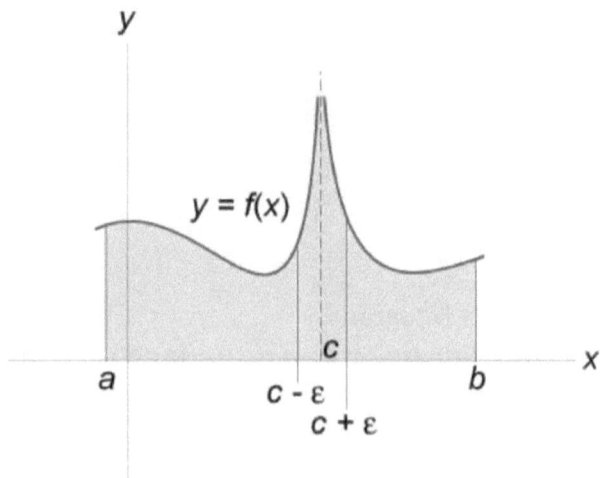

Figure 188.

9.10 Double Integral

Functions of two variables: $f(x,y)$, $f(u,v)$, ...

Double integrals: $\iint_R f(x,y)dxdy$, $\iint_R g(x,y)dxdy$, ...

Riemann sum: $\sum_{i=1}^{m}\sum_{j=1}^{n} f(u_i, v_j)\Delta x_i \Delta y_j$

Small changes: Δx_i, Δy_j

Regions of integration: R, S

Polar coordinates: r, θ

Area: A
Surface area: S
Volume of a solid: V
Mass of a lamina: m
Density: $\rho(x,y)$
First moments: M_x, M_y
Moments of inertia: I_x, I_y, I_0
Charge of a plate: Q
Charge density: $\sigma(x,y)$
Coordinates of center of mass: $\bar{x}$, $\bar{y}$
Average of a function: μ

1078. Definition of Double Integral
The double integral over a rectangle $[a, b] \times [c, d]$ is defined to be

$$\iint\limits_{[a,b]\times[c,d]} f(x,y) dA = \lim_{\substack{\max \Delta x_i \to 0 \\ \max \Delta y_j \to 0}} \sum_{i=1}^{m} \sum_{j=1}^{n} f(u_i, v_j) \Delta x_i \Delta y_j,$$

where (u_i, v_j) is some point in the rectangle $(x_{i-1}, x_i) \times (y_{j-1}, y_j)$, and $\Delta x_i = x_i - x_{i-1}$, $\Delta y_j = y_j - y_{j-1}$.

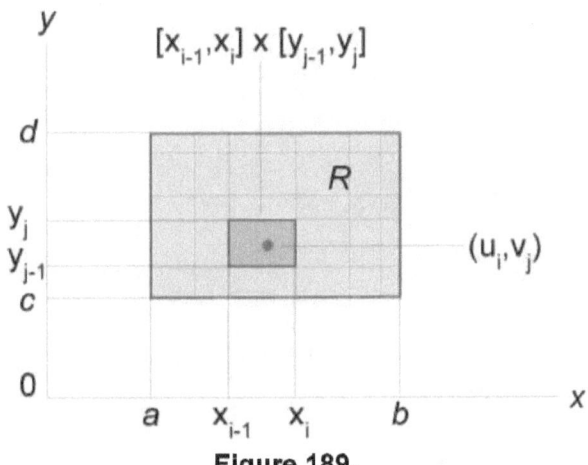

Figure 189.

The double integral over a general region R is

$$\iint_R f(x,y)\,dA = \iint_{[a,b]\times[c,d]} g(x,y)\,dA,$$

where rectangle $[a, b]\times[c, d]$ contains R, $g(x,y)=f(x,y)$ if $f(x,y)$ is in R and $g(x,y)=0$ otherwise.

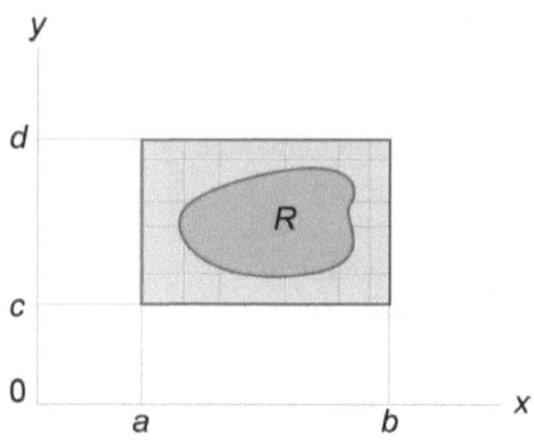

Figure 190.

1079. $\iint_R [f(x,y)+g(x,y)]\,dA = \iint_R f(x,y)\,dA + \iint_R g(x,y)\,dA$

1080. $\iint_R [f(x,y)-g(x,y)]\,dA = \iint_R f(x,y)\,dA - \iint_R g(x,y)\,dA$

1081. $\iint_R kf(x,y)\,dA = k\iint_R f(x,y)\,dA,$

where k is a constant.

1082. If $f(x,y)\le g(x,y)$ on R, then $\iint_R f(x,y)\,dA \le \iint_R g(x,y)\,dA$.

1083. If $f(x,y)\ge 0$ on R and $S\subset R$, then

$$\iint_S f(x,y)dA \le \iint_R f(x,y)dA.$$

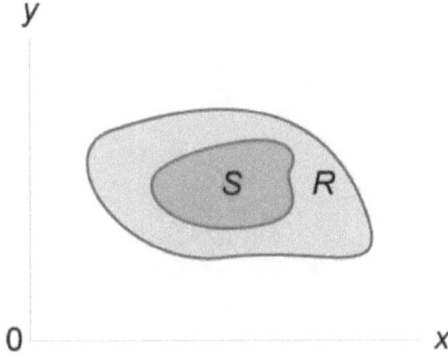

Figure 191.

1084. If $f(x,y) \ge 0$ on R and R and S are non-overlapping regions, then $\iint_{R \cup S} f(x,y)dA = \iint_R f(x,y)dA + \iint_S f(x,y)dA$.
Here $R \cup S$ is the union of the regions R and S.

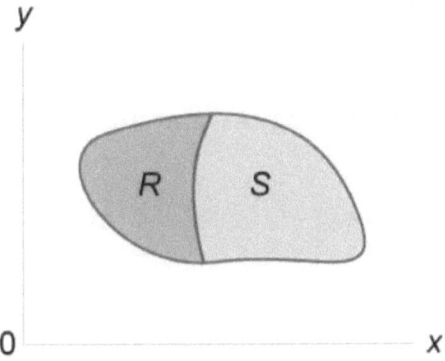

Figure 192.

1085. Iterated Integrals and Fubini's Theorem

$$\iint_R f(x,y)\,dA = \int_a^b \int_{p(x)}^{q(x)} f(x,y)\,dy\,dx$$

for a region of type I,
$R = \{(x,y) \mid a \leq x \leq b, p(x) \leq y \leq q(x)\}$.

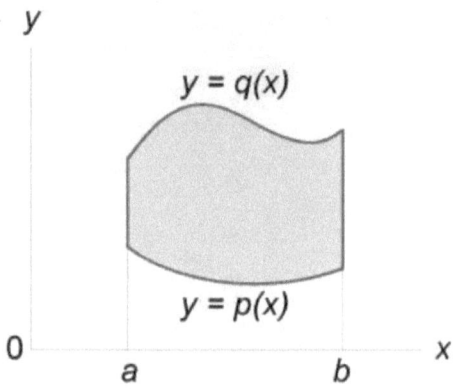

Figure 193.

$$\iint_R f(x,y)\,dA = \int_c^d \int_{u(y)}^{v(y)} f(x,y)\,dx\,dy$$

for a region of type II,
$R = \{(x,y) \mid u(y) \leq x \leq v(y), c \leq y \leq d\}$.

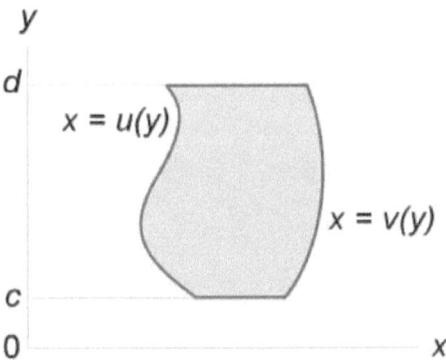

Figure 194.

1086. Double Integrals over Rectangular Regions

If R is the rectangular region $[a,b] \times [c,d]$, then

$$\iint_R f(x,y)dxdy = \int_a^b \left(\int_c^d f(x,y)dy \right) dx = \int_c^d \left(\int_a^b f(x,y)dx \right) dy.$$

In the special case where the integrand $f(x,y)$ can be written as $g(x)h(y)$ we have

$$\iint_R f(x,y)dxdy = \iint_R g(x)h(y)dxdy = \left(\int_a^b g(x)dx \right) \left(\int_c^d h(y)dy \right).$$

1087. Change of Variables

$$\iint_R f(x,y)dxdy = \iint_S f[x(u,v),y(u,v)] \left| \frac{\partial(x,y)}{\partial(u,v)} \right| dudv,$$

where $\left| \dfrac{\partial(x,y)}{\partial(u,v)} \right| = \begin{vmatrix} \dfrac{\partial x}{\partial u} & \dfrac{\partial x}{\partial v} \\ \dfrac{\partial y}{\partial u} & \dfrac{\partial y}{\partial v} \end{vmatrix} \neq 0$ is the jacobian of the transformations $(x,y) \to (u,v)$, and S is the pullback of R which can be computed by $x = x(u,v)$, $y = y(u,v)$ into the definition of R.

1088. Polar Coordinates
$x = r\cos\theta$, $y = r\sin\theta$.

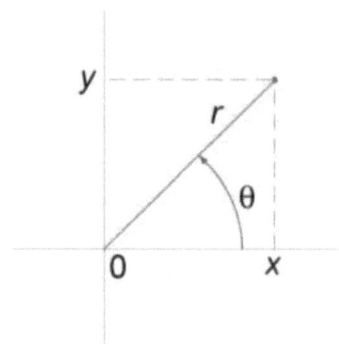

Figure 195.

1089. Double Integrals in Polar Coordinates

The Differential dxdy for Polar Coordinates is
$$dxdy = \left|\frac{\partial(x,y)}{\partial(r,\theta)}\right| drd\theta = rdrd\theta.$$

Let the region R is determined as follows:
$0 \le g(\theta) \le r \le h(\theta)$, $\alpha \le \theta \le \beta$, where $\beta - \alpha \le 2\pi$.
Then
$$\iint_R f(x,y)dxdy = \int_\alpha^\beta \int_{g(\theta)}^{h(\theta)} f(r\cos\theta, r\sin\theta)rdrd\theta.$$

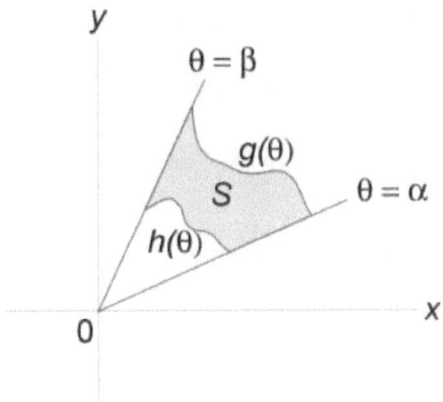

Figure 196.

If the region R is the polar rectangle given by $0 \leq a \leq r \leq b$, $\alpha \leq \theta \leq \beta$, where $\beta - \alpha \leq 2\pi$, then

$$\iint_R f(x,y)dxdy = \int_\alpha^\beta \int_a^b f(r\cos\theta, r\sin\theta)rdrd\theta.$$

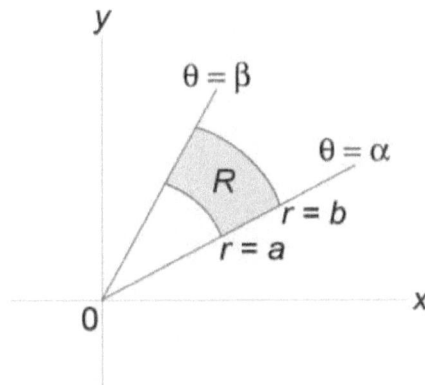

Figure 197.

1090. Area of a Region

$$A = \int_a^b \int_{g(x)}^{f(x)} dy\, dx \quad \text{(for a type I region)}.$$

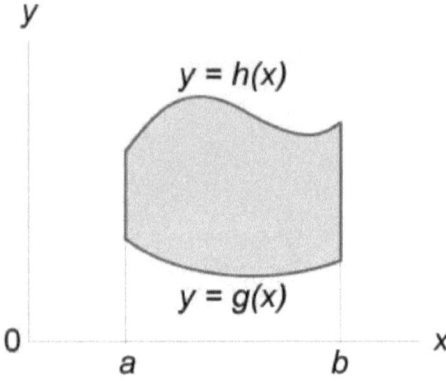

Figure 198.

$$A = \int_c^d \int_{p(y)}^{q(y)} dx\, dy \quad \text{(for a type II region)}.$$

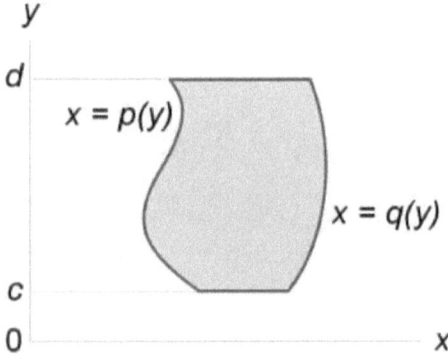

Figure 199.

1091. Volume of a Solid
$$V = \iint_R f(x,y)dA.$$

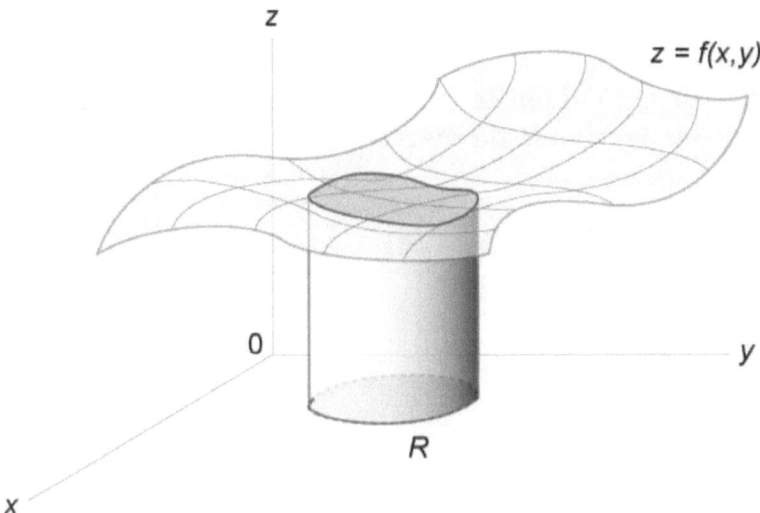

Figure 200.

If R is a type I region bounded by $x=a$, $x=b$, $y=h(x)$, $y=g(x)$, then

$$V = \iint_R f(x,y)dA = \int_a^b \int_{h(x)}^{g(x)} f(x,y)dydx.$$

If R is a type II region bounded by $y=c$, $y=d$, $x=q(y)$, $x=p(y)$, then

$$V = \iint_R f(x,y)dA = \int_c^d \int_{p(y)}^{q(y)} f(x,y)dxdy.$$

If $f(x,y) \geq g(x,y)$ over a region R, then the volume of the solid between $z_1 = f(x,y)$ and $z_2 = g(x,y)$ over R is given by

$$V = \iint_R [f(x,y) - g(x,y)] dA.$$

1092. Area and Volume in Polar Coordinates

If S is a region in the xy-plane bounded by $\theta = \alpha$, $\theta = \beta$, $r = h(\theta)$, $r = g(\theta)$, then

$$A = \iint_S dA = \int_\alpha^\beta \int_{h(\theta)}^{g(\theta)} r \, dr \, d\theta,$$

$$V = \iint_S f(r, \theta) r \, dr \, d\theta.$$

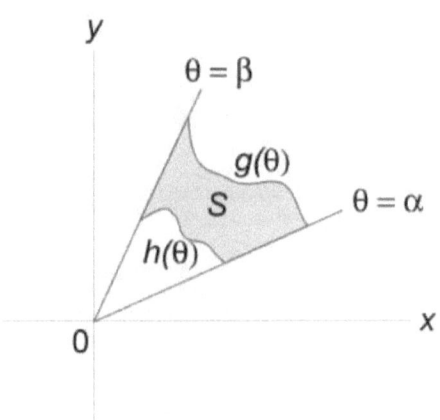

Figure 201.

1093. Surface Area

$$S = \iint_R \sqrt{1 + \left(\frac{\partial z}{\partial x}\right)^2 + \left(\frac{\partial z}{\partial y}\right)^2} \, dx \, dy$$

CHAPTER 9. INTEGRAL CALCULUS

1094. Mass of a Lamina
$$m = \iint_R \rho(x,y)\,dA,$$
where the lamina occupies a region R and its density at a point (x,y) is $\rho(x,y)$.

1095. Moments
The moment of the lamina about the x-axis is given by formula
$$M_x = \iint_R y\rho(x,y)\,dA.$$

The moment of the lamina about the y-axis is
$$M_y = \iint_R x\rho(x,y)\,dA.$$

The moment of inertia about the x-axis is
$$I_x = \iint_R y^2 \rho(x,y)\,dA.$$

The moment of inertia about the y-axis is
$$I_y = \iint_R x^2 \rho(x,y)\,dA.$$

The polar moment of inertia is
$$I_0 = \iint_R (x^2 + y^2)\rho(x,y)\,dA.$$

1096. Center of Mass
$$\bar{x} = \frac{M_y}{m} = \frac{1}{m}\iint_R x\rho(x,y)\,dA = \frac{\iint_R x\rho(x,y)\,dA}{\iint_R \rho(x,y)\,dA},$$

$$y = \frac{M_x}{m} = \frac{1}{m}\iint_R y\rho(x,y)dA = \frac{\iint_R y\rho(x,y)dA}{\iint_R \rho(x,y)dA}.$$

1097. Charge of a Plate

$$Q = \iint_R \sigma(x,y)dA,$$

where electrical charge is distributed over a region R and its charge density at a point (x,y) is $\sigma(x,y)$.

1098. Average of a Function

$$\mu = \frac{1}{S}\iint_R f(x,y)dA,$$

where $S = \iint_R dA$.

9.11 Triple Integral

Functions of three variables: $f(x,y,z)$, $g(x,y,z)$, ...

Triple integrals: $\iiint_G f(x,y,z)dV$, $\iiint_G g(x,y,z)dV$, ...

Riemann sum: $\sum_{i=1}^{m}\sum_{j=1}^{n}\sum_{k=1}^{p} f(u_i, v_j, w_k)\Delta x_i \Delta y_j \Delta z_k$

Small changes: Δx_i, Δy_j, Δz_k

Limits of integration: a, b, c, d, r, s

Regions of integration: G, T, S

Cylindrical coordinates: r, θ, z

Spherical coordinates: r, θ, φ

Volume of a solid: V

Mass of a solid: m
Density: $\mu(x,y,z)$
Coordinates of center of mass: $\bar{x}, \bar{y}, \bar{z}$
First moments: M_{xy}, M_{yz}, M_{xz}
Moments of inertia: $I_{xy}, I_{yz}, I_{xz}, I_x, I_y, I_z, I_0$

1099. Definition of Triple Integral
The triple integral over a parallelepiped $[a,b] \times [c,d] \times [r,s]$ is defined to be
$$\iiint_{[a,b]\times[c,d]\times[r,s]} f(x,y,z)dV = \lim_{\substack{\max \Delta x_i \to 0 \\ \max \Delta y_j \to 0 \\ \max \Delta z_k \to 0}} \sum_{i=1}^{m} \sum_{j=1}^{n} \sum_{k=1}^{p} f(u_i, v_j, w_k) \Delta x_i \Delta y_j \Delta z_k,$$
where (u_i, v_j, w_k) is some point in the parallelepiped $(x_{i-1}, x_i) \times (y_{j-1}, y_j) \times (z_{k-1}, z_k)$, and $\Delta x_i = x_i - x_{i-1}$, $\Delta y_j = y_j - y_{j-1}$, $\Delta z_k = z_k - z_{k-1}$.

1100. $\iiint_G [f(x,y,z) + g(x,y,z)]dV = \iiint_G f(x,y,z)dV + \iiint_G g(x,y,z)dV$

1101. $\iiint_G [f(x,y,z) - g(x,y,z)]dV = \iiint_G f(x,y,z)dV - \iiint_G g(x,y,z)dV$

1102. $\iiint_G kf(x,y,z)dV = k\iiint_G f(x,y,z)dV,$
where k is a constant.

1103. If $f(x,y,z) \geq 0$ and G and T are nonoverlapping basic regions, then
$$\iiint_{G \cup T} f(x,y,z)dV = \iiint_G f(x,y,z)dV + \iiint_T f(x,y,z)dV.$$
Here $G \cup T$ is the union of the regions G and T.

1104. Evaluation of Triple Integrals by Repeated Integrals

If the solid G is the set of points (x,y,z) such that $(x,y) \in R$, $\chi_1(x,y) \leq z \leq \chi_2(x,y)$, then

$$\iiint_G f(x,y,z)dxdydz = \iint_R \left[\int_{\chi_1(x,y)}^{\chi_2(x,y)} f(x,y,z)dz \right] dxdy,$$

where R is projection of G onto the xy-plane.

If the solid G is the set of points (x,y,z) such that $a \leq x \leq b$, $\varphi_1(x) \leq y \leq \varphi_2(x)$, $\chi_1(x,y) \leq z \leq \chi_2(x,y)$, then

$$\iiint_G f(x,y,z)dxdydz = \int_a^b \left[\int_{\varphi_1(x)}^{\varphi_2(x)} \left(\int_{\chi_1(x,y)}^{\chi_2(x,y)} f(x,y,z)dz \right) dy \right] dx$$

1105. Triple Integrals over Parallelepiped

If G is a parallelepiped $[a, b] \times [c, d] \times [r, s]$, then

$$\iiint_G f(x,y,z)dxdydz = \int_a^b \left[\int_c^d \left(\int_r^s f(x,y,z)dz \right) dy \right] dx.$$

In the special case where the integrand $f(x,y,z)$ can be written as $g(x)h(y)k(z)$ we have

$$\iiint_G f(x,y,z)dxdydz = \left(\int_a^b g(x)dx \right) \left(\int_c^d h(y)dy \right) \left(\int_r^s k(z)dz \right).$$

1106. Change of Variables

$$\iiint_G f(x,y,z)dxdydz =$$

$$= \iiint_S f[x(u,v,w), y(u,v,w), z(u,v,w)] \left| \frac{\partial(x,y,z)}{\partial(u,v,w)} \right| dxdydz,$$

where $\left|\dfrac{\partial(x,y,z)}{\partial(u,v,w)}\right| = \begin{vmatrix} \dfrac{\partial x}{\partial u} & \dfrac{\partial x}{\partial v} & \dfrac{\partial x}{\partial w} \\ \dfrac{\partial y}{\partial u} & \dfrac{\partial y}{\partial v} & \dfrac{\partial y}{\partial w} \\ \dfrac{\partial z}{\partial u} & \dfrac{\partial z}{\partial v} & \dfrac{\partial z}{\partial w} \end{vmatrix} \neq 0$ is the jacobian of the transformations $(x,y,z) \to (u,v,w)$, and S is the pullback of G which can be computed by $x = x(u,v,w)$, $y = y(u,v,w)$, $z = z(u,v,w)$ into the definition of G.

1107. Triple Integrals in Cylindrical Coordinates
The differential dxdydz for cylindrical coordinates is

$$dxdydz = \left|\dfrac{\partial(x,y,z)}{\partial(r,\theta,z)}\right| drd\theta dz = rdrd\theta dz.$$

Let the solid G is determined as follows:
$(x,y) \in R$, $\chi_1(x,y) \leq z \leq \chi_2(x,y)$,
where R is projection of G onto the xy-plane. Then

$$\iiint_G f(x,y,z)dxdydz = \iiint_S f(r\cos\theta, r\sin\theta, z) rdrd\theta dz$$

$$= \iint_{R(r,\theta)} \left[\int_{\chi_1(r\cos\theta, r\sin\theta)}^{\chi_2(r\cos\theta, r\sin\theta)} f(r\cos\theta, r\sin\theta, z) dz\right] rdrd\theta.$$

Here S is the pullback of G in cylindrical coordinates.

1108. Triple Integrals in Spherical Coordinates
The Differential dxdydz for Spherical Coordinates is

$$dxdydz = \left|\dfrac{\partial(x,y,z)}{\partial(r,\theta,\varphi)}\right| drd\theta d\varphi = r^2 \sin\theta drd\theta d\varphi$$

$$\iiint_G f(x,y,z)dxdydz =$$

$$= \iiint\limits_{S} f(r\sin\theta\cos\varphi, r\sin\theta\sin\varphi, r\cos\theta) r^2 \sin\theta \, dr \, d\theta \, d\varphi,$$

where the solid S is the pullback of G in spherical coordinates. The angle θ ranges from 0 to 2π, the angle φ ranges from 0 to π.

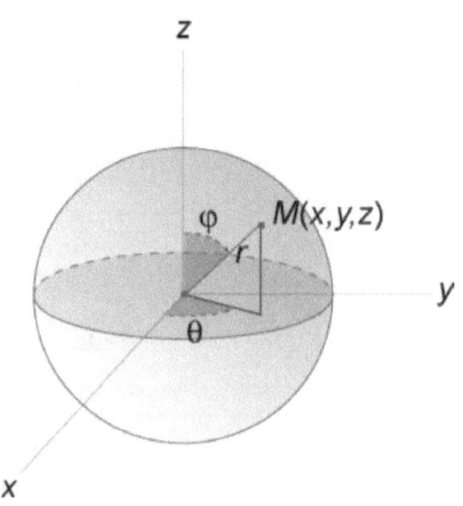

Figure 202.

1109. Volume of a Solid
$$V = \iiint\limits_{G} dxdydz$$

1110. Volume in Cylindrical Coordinates
$$V = \iiint\limits_{S(r,\theta,z)} r \, dr \, d\theta \, dz$$

1111. Volume in Spherical Coordinates
$$V = \iiint\limits_{S(r,\theta,\varphi)} r^2 \sin\theta \, dr \, d\theta \, d\varphi$$

1112. Mass of a Solid
$$m = \iiint_G \mu(x,y,z)\,dV,$$
where the solid occupies a region G and its density at a point (x,y,z) is $\mu(x,y,z)$.

1113. Center of Mass of a Solid
$$\bar{x} = \frac{M_{yz}}{m}, \quad \bar{y} = \frac{M_{xz}}{m}, \quad \bar{z} = \frac{M_{xy}}{m},$$
where
$$M_{yz} = \iiint_G x\mu(x,y,z)\,dV,$$
$$M_{xz} = \iiint_G y\mu(x,y,z)\,dV,$$
$$M_{xy} = \iiint_G z\mu(x,y,z)\,dV$$
are the first moments about the coordinate planes $x=0$, $y=0$, $z=0$, respectively, $\mu(x,y,z)$ is the density function.

1114. Moments of Inertia about the xy-plane (or $z=0$), yz-plane ($x=0$), and xz-plane ($y=0$)
$$I_{xy} = \iiint_G z^2\mu(x,y,z)\,dV,$$
$$I_{yz} = \iiint_G x^2\mu(x,y,z)\,dV,$$
$$I_{xz} = \iiint_G y^2\mu(x,y,z)\,dV.$$

1115. Moments of Inertia about the x-axis, y-axis, and z-axis
$$I_x = I_{xy} + I_{xz} = \iiint_G (z^2+y^2)\mu(x,y,z)\,dV,$$
$$I_y = I_{xy} + I_{yz} = \iiint_G (z^2+x^2)\mu(x,y,z)\,dV,$$

$$I_z = I_{xz} + I_{yz} = \iiint\limits_G (y^2 + x^2)\mu(x,y,z)dV.$$

1116. Polar Moment of Inertia
$$I_0 = I_{xy} + I_{yz} + I_{xz} = \iiint\limits_G (x^2 + y^2 + z^2)\mu(x,y,z)dV$$

9.12 Line Integral

Scalar functions: $F(x,y,z)$, $F(x,y)$, $f(x)$
Scalar potential: $u(x,y,z)$
Curves: C, C_1, C_2
Limits of integrations: a, b, α, β
Parameters: t, s
Polar coordinates: r, θ
Vector field: $\vec{F}(P,Q,R)$
Position vector: $\vec{r}(s)$
Unit vectors: $\vec{i}, \vec{j}, \vec{k}, \vec{\tau}$
Area of region: S
Length of a curve: L
Mass of a wire: m
Density: $\rho(x,y,z)$, $\rho(x,y)$
Coordinates of center of mass: $\bar{x}, \bar{y}, \bar{z}$
First moments: M_{xy}, M_{yz}, M_{xz}
Moments of inertia: I_x, I_y, I_z
Volume of a solid: V
Work: W
Magnetic field: $\vec{B}$
Current: I
Electromotive force: ε
Magnetic flux: ψ

CHAPTER 9. INTEGRAL CALCULUS

1117. Line Integral of a Scalar Function
Let a curve C be given by the vector function $\bar{r} = \bar{r}(s)$, $0 \le s \le S$, and a scalar function F is defined over the curve C. Then

$$\int_0^S F(\dot{r}(s))ds = \int_C F(x,y,z)ds = \int_C F ds,$$

where ds is the arc length differential.

1118. $\displaystyle\int_{C_1 \cup C_2} F\,ds = \int_{C_1} F\,ds + \int_{C_2} F\,ds$

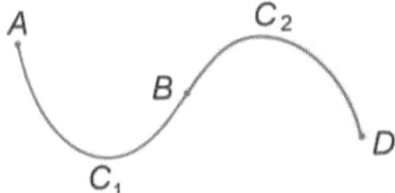

Figure 203.

1119. If the smooth curve C is parametrized by $\bar{r} = \bar{r}(t)$, $\alpha \le t \le \beta$, then

$$\int_C F(x,y,z)ds = \int_\alpha^\beta F(x(t),y(t),z(t))\sqrt{(x'(t))^2 + (y'(t))^2 + (z'(t))^2}\,dt.$$

1120. If C is a smooth curve in the xy-plane given by the equation $y = f(x)$, $a \le x \le b$, then

$$\int_C F(x,y)ds = \int_a^b F(x,f(x))\sqrt{1+(f'(x))^2}\,dx.$$

1121. Line Integral of Scalar Function in Polar Coordinates

$$\int_C F(x,y)ds = \int_\alpha^\beta F(r\cos\theta, r\sin\theta)\sqrt{r^2 + \left(\frac{dr}{d\theta}\right)^2}\,d\theta,$$

where the curve C is defined by the polar function r(θ).

1122. Line Integral of Vector Field
Let a curve C be defined by the vector function $\vec{r} = \vec{r}(s)$, $0 \le s \le S$. Then
$$\frac{d\vec{r}}{ds} = \vec{\tau} = (\cos\alpha, \cos\beta, \cos\gamma)$$
is the unit vector of the tangent line to this curve.

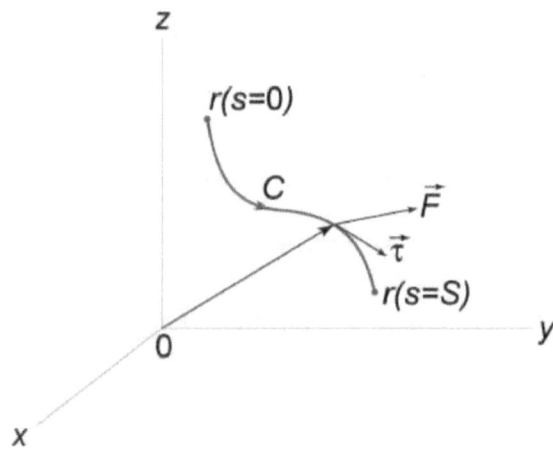

Figure 204.

Let a vector field $\vec{F}(P,Q,R)$ is defined over the curve C. Then the line integral of the vector field $\vec{F}$ along the curve C is

$$\int_C P\,dx + Q\,dy + R\,dz = \int_0^S (P\cos\alpha + Q\cos\beta + R\cos\gamma)\,ds.$$

1123. Properties of Line Integrals of Vector Fields

$$\int_{-C}(\vec{F}\cdot d\vec{r}) = -\int_{C}(\vec{F}\cdot d\vec{r}),$$

where -C denote the curve with the opposite orientation.

$$\int_{C}(\vec{F}\cdot d\vec{r}) = \int_{C_1 \cup C_2}(\vec{F}\cdot d\vec{r}) = \int_{C_1}(\vec{F}\cdot d\vec{r}) + \int_{C_2}(\vec{F}\cdot d\vec{r}),$$

where C is the union of the curves C_1 and C_2.

1124.
If the curve C is parameterized by $\vec{r}(t) = \langle x(t), y(t), z(t) \rangle$, $\alpha \le t \le \beta$, then

$$\int_C Pdx + Qdy + Rdz =$$

$$= \int_{\alpha}^{\beta} \left(P(x(t),y(t),z(t))\frac{dx}{dt} + Q(x(t),y(t),z(t))\frac{dy}{dt} + R(x(t),y(t),z(t))\frac{dz}{dt} \right) dt$$

1125.
If C lies in the xy-plane and given by the equation $y = f(x)$, then

$$\int_C Pdx + Qdy = \int_a^b \left(P(x,f(x)) + Q(x,f(x))\frac{df}{dx} \right) dx.$$

1126. Green's Theorem

$$\iint_R \left(\frac{\partial Q}{\partial x} - \frac{\partial P}{\partial y} \right) dxdy = \oint_C Pdx + Qdy,$$

where $\vec{F} = P(x,y)\vec{i} + Q(x,y)\vec{j}$ is a continuous vector function with continuous first partial derivatives $\frac{\partial P}{\partial y}$, $\frac{\partial Q}{\partial x}$ in a some domain R, which is bounded by a closed, piecewise smooth curve C.

CHAPTER 9. INTEGRAL CALCULUS

1127. Area of a Region R Bounded by the Curve C
$$S = \iint_R dx\,dy = \frac{1}{2}\oint_C x\,dy - y\,dx$$

1128. Path Independence of Line Integrals
The line integral of a vector function $\vec{F} = P\vec{i} + Q\vec{j} + R\vec{k}$ is said to be path independent, if and only if P, Q, and R are continuous in a domain D, and if there exists some scalar function $u = u(x,y,z)$ (a scalar potential) in D such that
$\vec{F} = \text{grad } u$, or $\frac{\partial u}{\partial x} = P$, $\frac{\partial u}{\partial y} = Q$, $\frac{\partial u}{\partial z} = R$.

Then
$$\int_C \vec{F}(\vec{r}) \cdot d\vec{r} = \int_C P\,dx + Q\,dy + R\,dz = u(B) - u(A).$$

1129. Test for a Conservative Field
A vector field of the form $\vec{F} = \text{grad } u$ is called a conservative field. The line integral of a vector function $\vec{F} = P\vec{i} + Q\vec{j} + R\vec{k}$ is path independent if and only if

$$\text{curl } \vec{F} = \begin{vmatrix} \vec{i} & \vec{j} & \vec{k} \\ \frac{\partial}{\partial x} & \frac{\partial}{\partial y} & \frac{\partial}{\partial z} \\ P & Q & R \end{vmatrix} = \vec{0}.$$

If the line integral is taken in xy-plane so that
$$\int_C P\,dx + Q\,dy = u(B) - u(A),$$

then the test for determining if a vector field is conservative can be written in the form
$$\frac{\partial P}{\partial y} = \frac{\partial Q}{\partial x}.$$

1130. Length of a Curve

$$L = \int_C ds = \int_\alpha^\beta \left|\frac{d\vec{r}}{dt}(t)\right| dt = \int_\alpha^\beta \sqrt{\left(\frac{dx}{dt}\right)^2 + \left(\frac{dy}{dt}\right)^2 + \left(\frac{dz}{dt}\right)^2}\, dt,$$

where C ia a piecewise smooth curve described by the position vector $\vec{r}(t)$, $\alpha \leq t \leq \beta$.

If the curve C is two-dimensional, then

$$L = \int_C ds = \int_\alpha^\beta \left|\frac{d\vec{r}}{dt}(t)\right| dt = \int_\alpha^\beta \sqrt{\left(\frac{dx}{dt}\right)^2 + \left(\frac{dy}{dt}\right)^2}\, dt.$$

If the curve C is the graph of a function $y = f(x)$ in the xy-plane $(a \leq x \leq b)$, then

$$L = \int_a^b \sqrt{1 + \left(\frac{dy}{dx}\right)^2}\, dx.$$

1131. Length of a Curve in Polar Coordinates

$$L = \int_\alpha^\beta \sqrt{\left(\frac{dr}{d\theta}\right)^2 + r^2}\, d\theta,$$

where the curve C is given by the equation $r = r(\theta)$, $\alpha \leq \theta \leq \beta$ in polar coordinates.

1132. Mass of a Wire

$$m = \int_C \rho(x, y, z)\, ds,$$

where $\rho(x, y, z)$ is the mass per unit length of the wire.

If C is a curve parametrized by the vector function $\vec{r}(t) = \langle x(t), y(t), z(t) \rangle$, then the mass can be computed by the formula

$$m = \int_\alpha^\beta \rho(x(t), y(t), z(t)) \sqrt{\left(\frac{dx}{dt}\right)^2 + \left(\frac{dy}{dt}\right)^2 + \left(\frac{dz}{dt}\right)^2}\, dt.$$

If C is a curve in xy-plane, then the mass of the wire is given by
$$m = \int_C \rho(x, y)\, ds,$$

or

$$m = \int_\alpha^\beta \rho(x(t), y(t)) \sqrt{\left(\frac{dx}{dt}\right)^2 + \left(\frac{dy}{dt}\right)^2}\, dt \text{ (in parametric form)}.$$

1133. Center of Mass of a Wire

$$\bar{x} = \frac{M_{yz}}{m},\ \bar{y} = \frac{M_{xz}}{m},\ \bar{z} = \frac{M_{xy}}{m},$$

where

$$M_{yz} = \int_C x\rho(x, y, z)\, ds,$$

$$M_{xz} = \int_C y\rho(x, y, z)\, ds,$$

$$M_{xy} = \int_C z\rho(x, y, z)\, ds.$$

1134. Moments of Inertia

The moments of inertia about the x-axis, y-axis, and z-axis are given by the formulas

$$I_x = \int_C (y^2 + z^2)\rho(x, y, z)\, ds,$$

$$I_y = \int_C (x^2 + z^2)\rho(x, y, z)\, ds,$$

$$I_z = \int_C (x^2 + y^2)\rho(x, y, z)\, ds.$$

1135. Area of a Region Bounded by a Closed Curve

$$S = \oint_C x\,dy = -\oint_C y\,dx = \frac{1}{2}\oint_C x\,dy - y\,dx.$$

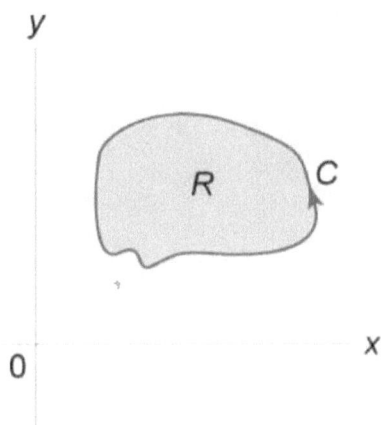

Figure 205.

If the closed curve C is given in parametric form $\vec{r}(t) = \langle x(t), y(t) \rangle$, then the area can be calculated by the formula

$$S = \int_\alpha^\beta x(t)\frac{dy}{dt}dt = -\int_\alpha^\beta y(t)\frac{dx}{dt}dt = \frac{1}{2}\int_\alpha^\beta \left(x(t)\frac{dy}{dt} - y(t)\frac{dx}{dt} \right) dt.$$

1136. Volume of a Solid Formed by Rotating a Closed Curve about the x-axis

$$V = -\pi \oint_C y^2\,dx = -2\pi \oint_C xy\,dy = -\frac{\pi}{2}\oint_C 2xy\,dy + y^2\,dx.$$

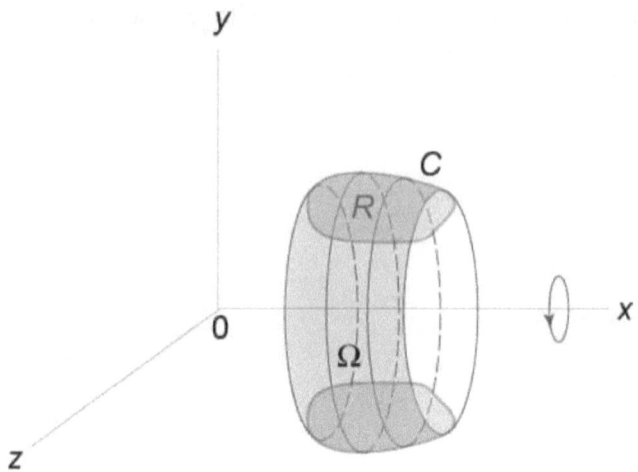

Figure 206.

1137. Work

Work done by a force $\vec{F}$ on an object moving along a curve C is given by the line integral

$$W = \int_C \vec{F} \cdot d\vec{r},$$

where $\vec{F}$ is the vector force field acting on the object, $d\vec{r}$ is the unit tangent vector.

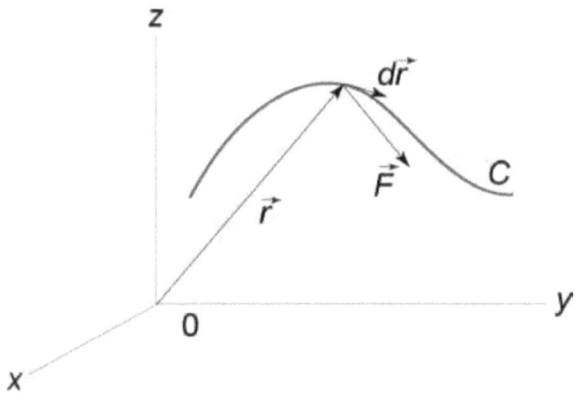

Figure 207.

If the object is moved along a curve C in the xy-plane, then
$$W = \int_C \vec{F} \cdot d\vec{r} = \int_C P dx + Q dy,$$

If a path C is specified by a parameter t (t often means time), the formula for calculating work becomes
$$W = \int_\alpha^\beta \left[P(x(t), y(t), z(t)) \frac{dx}{dt} + Q(x(t), y(t), z(t)) \frac{dy}{dt} + R(x(t), y(t), z(t)) \frac{dz}{dt} \right] dt,$$
where t goes from α to β.

If a vector field $\vec{F}$ is conservative and $u(x,y,z)$ is a scalar potential of the field, then the work on an object moving from A to B can be found by the formula
$$W = u(B) - u(A).$$

1138. Ampere's Law
$$\oint_C \vec{B} \cdot d\vec{r} = \mu_0 I.$$

The line integral of a magnetic field $\vec{B}$ around a closed path C is equal to the total current I flowing through the area bounded by the path.

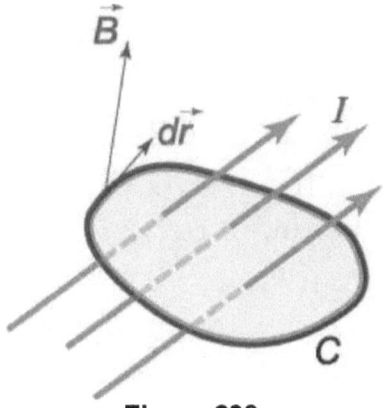

Figure 208.

1139. Faraday's Law

$$\varepsilon = \oint_C \vec{E} \cdot d\vec{r} = -\frac{d\psi}{dt}$$

The electromotive force (emf) ε induced around a closed loop C is equal to the rate of the change of magnetic flux ψ passing through the loop.

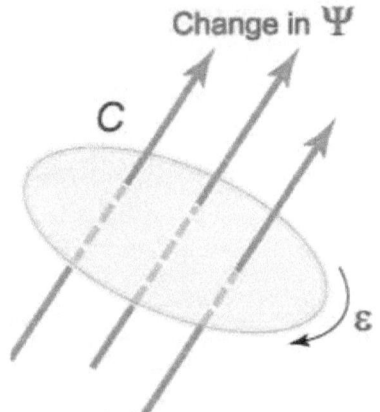

Figure 209.

9.13 Surface Integral

Scalar functions: $f(x,y,z)$, $z(x,y)$
Position vectors: $\vec{r}(u,v)$, $\vec{r}(x,y,z)$
Unit vectors: $\vec{i}$, $\vec{j}$, $\vec{k}$
Surface: S
Vector field: $\vec{F}(P,Q,R)$
Divergence of a vector field: $\text{div } \vec{F} = \nabla \cdot \vec{F}$

CHAPTER 9. INTEGRAL CALCULUS

Curl of a vector field: $\operatorname{curl} \vec{F} = \nabla \times \vec{F}$
Vector element of a surface: $d\vec{S}$
Normal to surface: $\vec{n}$
Surface area: A
Mass of a surface: m
Density: $\mu(x,y,z)$
Coordinates of center of mass: $\bar{x}, \bar{y}, \bar{z}$
First moments: M_{xy}, M_{yz}, M_{xz}
Moments of inertia: $I_{xy}, I_{yz}, I_{xz}, I_x, I_y, I_z$
Volume of a solid: V
Force: $\vec{F}$
Gravitational constant: G
Fluid velocity: $\vec{v}(\vec{r})$
Fluid density: ρ
Pressure: $p(\vec{r})$
Mass flux, electric flux: Φ
Surface charge: Q
Charge density: $\sigma(x,y)$
Magnitude of the electric field: $\vec{E}$

1140. Surface Integral of a Scalar Function
Let a surface S be given by the position vector
$\vec{r}(u,v) = x(u,v)\vec{i} + y(u,v)\vec{j} + z(u,v)\vec{k}$,
where (u,v) ranges over some domain $D(u,v)$ of the uv-plane.
The surface integral of a scalar function $f(x,y,z)$ over the surface S is defined as

$$\iint_S f(x,y,z) dS = \iint_{D(u,v)} f(x(u,v), y(u,v), z(u,v)) \left| \frac{\partial \vec{r}}{\partial u} \times \frac{\partial \vec{r}}{\partial v} \right| du\, dv,$$

where the partial derivatives $\dfrac{\partial \vec{r}}{\partial u}$ and $\dfrac{\partial \vec{r}}{\partial v}$ are given by

$$\frac{\partial \vec{r}}{\partial u} = \frac{\partial x}{\partial u}(u,v)\vec{i} + \frac{\partial y}{\partial u}(u,v)\vec{j} + \frac{\partial z}{\partial u}(u,v)\vec{k},$$

$$\frac{\partial \vec{r}}{\partial v} = \frac{\partial x}{\partial v}(u,v)\vec{i} + \frac{\partial y}{\partial v}(u,v)\vec{j} + \frac{\partial z}{\partial v}(u,v)\vec{k}$$

and $\dfrac{\partial \vec{r}}{\partial u} \times \dfrac{\partial \vec{r}}{\partial v}$ is the cross product.

1141. If the surface S is given by the equation $z = z(x,y)$ where $z(x,y)$ is a differentiable function in the domain $D(x,y)$, then

$$\iint_S f(x,y,z)\,dS = \iint_{D(x,y)} f(x,y,z(x,y))\sqrt{1 + \left(\frac{\partial z}{\partial x}\right)^2 + \left(\frac{\partial z}{\partial y}\right)^2}\,dx\,dy.$$

1142. Surface Integral of the Vector Field $\vec{F}$ over the Surface S
- If S is oriented outward, then

$$\iint_S \vec{F}(x,y,z)\cdot d\vec{S} = \iint_S \vec{F}(x,y,z)\cdot \vec{n}\,dS$$

$$= \iint_{D(u,v)} \vec{F}(x(u,v),y(u,v),z(u,v))\cdot \left[\frac{\partial \vec{r}}{\partial u} \times \frac{\partial \vec{r}}{\partial v}\right] du\,dv.$$

- If S is oriented inward, then

$$\iint_S \vec{F}(x,y,z)\cdot d\vec{S} = \iint_S \vec{F}(x,y,z)\cdot \vec{n}\,dS$$

$$= \iint_{D(u,v)} \vec{F}(x(u,v),y(u,v),z(u,v))\cdot \left[\frac{\partial \vec{r}}{\partial v} \times \frac{\partial \vec{r}}{\partial u}\right] du\,dv.$$

$d\vec{S} = \vec{n}\,dS$ is called the vector element of the surface. Dot means the scalar product of the appropriate vectors. The partial derivatives $\dfrac{\partial \vec{r}}{\partial u}$ and $\dfrac{\partial \vec{r}}{\partial v}$ are given by

$$\frac{\partial \vec{r}}{\partial u} = \frac{\partial x}{\partial u}(u,v)\cdot \vec{i} + \frac{\partial y}{\partial u}(u,v)\cdot \vec{j} + \frac{\partial z}{\partial u}(u,v)\cdot \vec{k},$$

$$\frac{\partial \vec{r}}{\partial v} = \frac{\partial x}{\partial v}(u,v)\cdot \vec{i} + \frac{\partial y}{\partial v}(u,v)\cdot \vec{j} + \frac{\partial z}{\partial v}(u,v)\cdot \vec{k}.$$

1143. If the surface S is given by the equation $z = z(x,y)$, where $z(x,y)$ is a differentiable function in the domain $D(x,y)$, then
- If S is oriented upward, i.e. the k-th component of the normal vector is positive, then

$$\iint_S \vec{F}(x,y,z)\cdot d\vec{S} = \iint_S \vec{F}(x,y,z)\cdot \vec{n}\, dS$$

$$= \iint_{D(x,y)} \vec{F}(x,y,z)\cdot \left(-\frac{\partial z}{\partial x}\vec{i} - \frac{\partial z}{\partial y}\vec{j} + \vec{k}\right) dxdy,$$

- If S is oriented downward, i.e. the k-th component of the normal vector is negative, then

$$\iint_S \vec{F}(x,y,z)\cdot d\vec{S} = \iint_S \vec{F}(x,y,z)\cdot \vec{n}\, dS$$

$$= \iint_{D(x,y)} \vec{F}(x,y,z)\cdot \left(\frac{\partial z}{\partial x}\vec{i} + \frac{\partial z}{\partial y}\vec{j} - \vec{k}\right) dxdy.$$

1144. $\iint_S (\vec{F}\cdot \vec{n})\, dS = \iint_S P\, dydz + Q\, dzdx + R\, dxdy$

$$= \iint_S (P\cos\alpha + Q\cos\beta + R\cos\gamma)\, dS,$$

where $P(x,y,z)$, $Q(x,y,z)$, $R(x,y,z)$ are the components of the vector field $\vec{F}$.

$\cos\alpha$, $\cos\beta$, $\cos\gamma$ are the angles between the outer unit normal vector $\vec{n}$ and the x-axis, y-axis, and z-axis, respectively.

CHAPTER 9. INTEGRAL CALCULUS

1145. If the surface S is given in parametric form by the vector $\vec{r}(x(u,v), y(u,v), z(u,v))$, then the latter formula can be written as

$$\iint_S (\vec{F}\cdot\vec{n})dS = \iint_S Pdydz + Qdzdx + Rdxdy = \iint_{D(u,v)} \begin{vmatrix} P & Q & R \\ \dfrac{\partial x}{\partial u} & \dfrac{\partial y}{\partial u} & \dfrac{\partial z}{\partial u} \\ \dfrac{\partial x}{\partial v} & \dfrac{\partial y}{\partial v} & \dfrac{\partial z}{\partial v} \end{vmatrix} dudv,$$

where (u,v) ranges over some domain $D(u,v)$ of the uv-plane.

1146. Divergence Theorem
$$\oiint_S \vec{F}\cdot d\vec{S} = \iiint_G (\nabla\cdot\vec{F})dV,$$

where

$$\vec{F}(x,y,z) = \langle P(x,y,z), Q(x,y,z), R(x,y,z)\rangle$$

is a vector field whose components P, Q, and R have continuous partial derivatives,

$$\nabla\cdot\vec{F} = \frac{\partial P}{\partial x} + \frac{\partial Q}{\partial y} + \frac{\partial R}{\partial z}$$

is the divergence of $\vec{F}$, also denoted div$\vec{F}$. The symbol $\oiint$ indicates that the surface integral is taken over a closed surface.

1147. Divergence Theorem in Coordinate Form
$$\oiint_S Pdydz + Qdxdz + Rdxdy = \iiint_G \left(\frac{\partial P}{\partial x} + \frac{\partial Q}{\partial y} + \frac{\partial R}{\partial z}\right)dxdydz.$$

1148. Stoke's Theorem
$$\oint_C \vec{F}\cdot d\vec{r} = \iint_S (\nabla\times\vec{F})\cdot d\vec{S},$$

where
$$\vec{F}(x,y,z) = \langle P(x,y,z), Q(x,y,z), R(x,y,z) \rangle$$
a vector field whose components P, Q, and R have continuous partial derivatives,

$$\nabla \times \vec{F} = \begin{vmatrix} \vec{i} & \vec{j} & \vec{k} \\ \frac{\partial}{\partial x} & \frac{\partial}{\partial x} & \frac{\partial}{\partial x} \\ P & Q & R \end{vmatrix} = \left(\frac{\partial R}{\partial y} - \frac{\partial Q}{\partial z} \right)\vec{i} + \left(\frac{\partial P}{\partial z} - \frac{\partial R}{\partial x} \right)\vec{j} + \left(\frac{\partial Q}{\partial x} - \frac{\partial P}{\partial y} \right)\vec{k}$$

is the curl of $\vec{F}$, also denoted curl $\vec{F}$.

The symbol $\oint$ indicates that the line integral is taken over a closed curve.

1149. Stoke's Theorem in Coordinate Form
$$\oint_C Pdx + Qdy + Rdz$$
$$= \iint_S \left(\frac{\partial R}{\partial y} - \frac{\partial Q}{\partial z} \right) dydz + \left(\frac{\partial P}{\partial z} - \frac{\partial R}{\partial x} \right) dzdx + \left(\frac{\partial Q}{\partial x} - \frac{\partial P}{\partial y} \right) dxdy$$

1150. Surface Area
$$A = \iint_S dS$$

1151. If the surface S is parameterized by the vector
$$\vec{r}(u,v) = x(u,v)\vec{i} + y(u,v)\vec{j} + z(u,v)\vec{k},$$
then the surface area is
$$A = \iint_{D(u,v)} \left| \frac{\partial \vec{r}}{\partial u} \times \frac{\partial \vec{r}}{\partial v} \right| dudv,$$
where $D(u,v)$ is the domain where the surface $\vec{r}(u,v)$ is defined.

CHAPTER 9. INTEGRAL CALCULUS

1152. If S is given explicitly by the function $z(x,y)$, then the surface area is

$$A = \iint_{D(x,y)} \sqrt{1+\left(\frac{\partial z}{\partial x}\right)^2 + \left(\frac{\partial z}{\partial y}\right)^2} \, dxdy,$$

where $D(x,y)$ is the projection of the surface S onto the xy-plane.

1153. Mass of a Surface

$$m = \iint_S \mu(x,y,z)dS,$$

where $\mu(x,y,z)$ is the mass per unit area (density function).

1154. Center of Mass of a Shell

$$\bar{x} = \frac{M_{yz}}{m}, \quad \bar{y} = \frac{M_{xz}}{m}, \quad \bar{z} = \frac{M_{xy}}{m},$$

where

$$M_{yz} = \iint_S x\mu(x,y,z)dS,$$

$$M_{xz} = \iint_S y\mu(x,y,z)dS,$$

$$M_{xy} = \iint_S z\mu(x,y,z)dS$$

are the first moments about the coordinate planes $x=0$, $y=0$, $z=0$, respectively. $\mu(x,y,z)$ is the density function.

1155. Moments of Inertia about the xy-plane (or $z=0$), yz-plane ($x=0$), and xz-plane ($y=0$)

$$I_{xy} = \iint_S z^2 \mu(x,y,z)dS,$$

$$I_{yz} = \iint_S x^2 \mu(x,y,z)dS,$$

$$I_{xz} = \iint_S y^2 \mu(x,y,z) dS.$$

1156. Moments of Inertia about the x-axis, y-axis, and z-axis
$$I_x = \iint_S (y^2 + z^2) \mu(x,y,z) dS,$$
$$I_y = \iint_S (x^2 + z^2) \mu(x,y,z) dS,$$
$$I_z = \iint_S (x^2 + y^2) \mu(x,y,z) dS.$$

1157. Volume of a Solid Bounded by a Closed Surface
$$V = \frac{1}{3} \left| \oiint_S x\,dydz + y\,dxdz + z\,dxdy \right|$$

1158. Gravitational Force
$$\vec{F} = Gm \iint_S \mu(x,y,z) \frac{\vec{r}}{r^3} dS,$$
where m is a mass at a point $\langle x_0, y_0, z_0 \rangle$ outside the surface,
$\vec{r} = \langle x - x_0, y - y_0, z - z_0 \rangle$,
$\mu(x,y,z)$ is the density function,
and G is gravitational constant.

1159. Pressure Force
$$\vec{F} = \iint_S p(\vec{r}) d\vec{S},$$
where the pressure $p(\vec{r})$ acts on the surface S given by the position vector $\vec{r}$.

1160. Fluid Flux (across the surface S)
$$\Phi = \oiint_S \vec{v}(\vec{r}) \cdot d\vec{S},$$

where $\vec{v}(\vec{r})$ is the fluid velocity.

1161. Mass Flux (across the surface S)
$$\Phi = \oiint_S \rho\vec{v}(\vec{r})\cdot d\vec{S},$$
where $\vec{F} = \rho\vec{v}$ is the vector field, ρ is the fluid density.

1162. Surface Charge
$$Q = \iint_S \sigma(x,y)dS,$$
where $\sigma(x,y)$ is the surface charge density.

1163. Gauss' Law
The electric flux through any closed surface is proportional to the charge Q enclosed by the surface
$$\Phi = \oiint_S \vec{E}\cdot d\vec{S} = \frac{Q}{\varepsilon_0},$$
where
Φ is the electric flux,
$\vec{E}$ is the magnitude of the electric field strength,
$\varepsilon_0 = 8{,}85\times 10^{-12}\,\dfrac{F}{m}$ is permittivity of free space.

Chapter 10
Differential Equations

Functions of one variable: y, p, q, u, g, h, G, H, r, z
Arguments (independent variables): x, y
Functions of two variables: $f(x,y)$, $M(x,y)$, $N(x,y)$

First order derivative: y', u', $\dot{y}$, $\dfrac{dy}{dt}$, ...

Second order derivatives: y'', $\ddot{y}$, $\dfrac{d^2 I}{dt^2}$, ...

Partial derivatives: $\dfrac{\partial u}{\partial t}$, $\dfrac{\partial^2 u}{\partial x^2}$, ...

Natural number: n
Particular solutions: y_1, y_p
Real numbers: k, t, C, C_1, C_2, p, q, α, β
Roots of the characteristic equations: λ_1, λ_2
Time: t
Temperature: T, S
Population function: $P(t)$
Mass of an object: m
Stiffness of a spring: k
Displacement of the mass from equilibrium: y
Amplitude of the displacement: A
Frequency: ω
Damping coefficient: γ
Phase angle of the displacement: δ
Angular displacement: θ
Pendulum length: L

Acceleration of gravity: g
Current: I
Resistance: R
Inductance: L
Capacitance: C

10.1 First Order Ordinary Differential Equations

1164. Linear Equations
$$\frac{dy}{dx} + p(x)y = q(x).$$

The general solution is
$$y = \frac{\int u(x)q(x)dx + C}{u(x)},$$
where
$$u(x) = \exp\left(\int p(x)dx\right).$$

1165. Separable Equations
$$\frac{dy}{dx} = f(x,y) = g(x)h(y)$$

The general solution is given by
$$\int \frac{dy}{h(y)} = \int g(x)dx + C,$$
or
$$H(y) = G(x) + C.$$

1166. Homogeneous Equations

The differential equation $\dfrac{dy}{dx} = f(x,y)$ is homogeneous, if the function $f(x,y)$ is homogeneous, that is $f(tx, ty) = f(x,y)$.

The substitution $z = \dfrac{y}{x}$ (then $y = zx$) leads to the separable equation

$$x\dfrac{dz}{dx} + z = f(1, z).$$

1167. Bernoulli Equation

$$\dfrac{dy}{dx} + p(x)y = q(x)y^n.$$

The substitution $z = y^{1-n}$ leads to the linear equation

$$\dfrac{dz}{dx} + (1-n)p(x)z = (1-n)q(x).$$

1168. Riccati Equation

$$\dfrac{dy}{dx} = p(x) + q(x)y + r(x)y^2$$

If a particular solution y_1 is known, then the general solution can be obtained with the help of substitution

$z = \dfrac{1}{y - y_1}$, which leads to the first order linear equation

$$\dfrac{dz}{dx} = -[q(x) + 2y_1 r(x)]z - r(x).$$

1169. Exact and Nonexact Equations

The equation
$$M(x,y)dx + N(x,y)dy = 0$$
is called exact if
$$\frac{\partial M}{\partial y} = \frac{\partial N}{\partial x},$$
and nonexact otherwise.

The general solution is
$$\int M(x,y)dx + \int N(x,y)dy = C.$$

1170. Radioactive Decay

$$\frac{dy}{dt} = -ky,$$
where $y(t)$ is the amount of radioactive element at time t, k is the rate of decay.

The solution is
$$y(t) = y_0 e^{-kt},$$ where $y_0 = y(0)$ is the initial amount.

1171. Newton's Law of Cooling

$$\frac{dT}{dt} = -k(T - S),$$
where $T(t)$ is the temperature of an object at time t, S is the temperature of the surrounding environment, k is a positive constant.

The solution is
$$T(t) = S + (T_0 - S)e^{-kt},$$
where $T_0 = T(0)$ is the initial temperature of the object at time $t = 0$.

1172. Population Dynamics (Logistic Model)
$$\frac{dP}{dt} = kP\left(1 - \frac{P}{M}\right),$$
where $P(t)$ is population at time t, k is a positive constant, M is a limiting size for the population.

The solution of the differential equation is
$$P(t) = \frac{MP_0}{P_0 + (M - P_0)e^{-kt}},$$ where $P_0 = P(0)$ is the initial population at time $t = 0$.

10.2 Second Order Ordinary Differential Equations

1173. Homogeneous Linear Equations with Constant Coefficients
$$y'' + py' + qy = 0.$$
The characteristic equation is
$$\lambda^2 + p\lambda + q = 0.$$

If λ_1 and λ_2 are distinct real roots of the characteristic equation, then the general solution is
$$y = C_1 e^{\lambda_1 x} + C_2 e^{\lambda_2 x},$$ where
C_1 and C_2 are integration constants.

If $\lambda_1 = \lambda_2 = -\frac{p}{2}$, then the general solution is
$$y = (C_1 + C_2 x)e^{-\frac{p}{2}x}.$$

If λ_1 and λ_2 are complex numbers:

$\lambda_1 = \alpha + \beta i$, $\lambda_2 = \alpha - \beta i$, where
$$\alpha = -\frac{p}{2}, \quad \beta = \frac{\sqrt{4q-p^2}}{2},$$
then the general solution is
$$y = e^{\alpha x}(C_1 \cos\beta x + C_2 \sin\beta x).$$

1174. Inhomogeneous Linear Equations with Constant Coefficients
$$y'' + py' + qy = f(x).$$

The general solution is given by
$y = y_p + y_h$, where
y_p is a particular solution of the inhomogeneous equation and y_h is the general solution of the associated homogeneous equation (see the previous topic 1173).

If the right side has the form
$$f(x) = e^{\alpha x}(P_1(x)\cos\beta x + P_1(x)\sin\beta x),$$
then the particular solution y_p is given by
$$y_p = x^k e^{\alpha x}(R_1(x)\cos\beta x + R_2(x)\sin\beta x),$$
where the polynomials $R_1(x)$ and $R_2(x)$ have to be found by using the method of undetermined coefficients.
- If $\alpha + \beta i$ is not a root of the characteristic equation, then the power $k = 0$,
- If $\alpha + \beta i$ is a simple root, then $k = 1$,
- If $\alpha + \beta i$ is a double root, then $k = 2$.

1175. Differential Equations with y Missing
$$y'' = f(x, y').$$
Set $u = y'$. Then the new equation satisfied by v is
$$u' = f(x, u),$$
which is a first order differential equation.

176. Differential Equations with x Missing
$y'' = f(y, y')$.

Set $u = y'$. Since
$$y'' = \frac{du}{dx} = \frac{du}{dy}\frac{dy}{dx} = u\frac{du}{dy},$$
we have
$$u\frac{du}{dy} = f(y, u),$$
which is a first order differential equation.

1177. Free Undamped Vibrations
The motion of a Mass on a Spring is described by the equation
$$m\ddot{y} + ky = 0,$$
where
m is the mass of the object,
k is the stiffness of the spring,
y is displacement of the mass from equilibrium.

The general solution is
$$y = A\cos(\omega_0 t - \delta),$$
where
A is the amplitude of the displacement,
ω_0 is the fundamental frequency, the period is $T = \frac{2\pi}{\omega_0}$,
δ is phase angle of the displacement.
This is an example of simple harmonic motion.

1178. Free Damped Vibrations
$m\ddot{y} + \gamma\dot{y} + ky = 0$, where
γ is the damping coefficient.
There are 3 cases for the general solution:

Case 1. $\gamma^2 > 4km$ (overdamped)
$$y(t) = Ae^{\lambda_1 t} + Be^{\lambda_2 t},$$
where
$$\lambda_1 = \frac{-\gamma - \sqrt{\gamma^2 - 4km}}{2m}, \quad \lambda_2 = \frac{-\gamma + \sqrt{\gamma^2 - 4km}}{2m}.$$

Case 2. $\gamma^2 = 4km$ (critically damped)
$$y(t) = (A + Bt)e^{\lambda t},$$
where
$$\lambda = -\frac{\gamma}{2m}.$$

Case 3. $\gamma^2 < 4km$ (underdamped)
$$y(t) = e^{-\frac{\gamma}{2m}t} A\cos(\omega t - \delta), \text{ where}$$
$$\omega = \sqrt{4km - \gamma^2}.$$

1179. Simple Pendulum
$$\frac{d^2\theta}{dt^2} + \frac{g}{L}\theta = 0,$$
where θ is the angular displacement, L is the pendulum length, g is the acceleration of gravity.

The general solution for small angles θ is
$$\theta(t) = \theta_{max} \sin\sqrt{\frac{g}{L}}t, \text{ the period is } T = 2\pi\sqrt{\frac{L}{g}}.$$

1180. RLC Circuit
$$L\frac{d^2I}{dt^2} + R\frac{dI}{dt} + \frac{1}{C}I = V'(t) = \omega E_0 \cos(\omega t),$$

where I is the current in an RLC circuit with an ac voltage source $V(t) = E_0 \sin(\omega t)$.

The general solution is
$$I(t) = C_1 e^{r_1 t} + C_2 e^{r_2 t} + A\sin(\omega t - \varphi),$$
where
$$r_{1,2} = \frac{-R \pm \sqrt{R^2 - \frac{4L}{C}}}{2L},$$

$$A = \frac{\omega E_0}{\sqrt{\left(L\omega^2 - \frac{1}{C}\right)^2 + R^2 \omega^2}},$$

$$\varphi = \arctan\left(\frac{L\omega}{R} - \frac{1}{RC\omega}\right),$$

C_1, C_2 are constants depending on initial conditions.

10.3. Some Partial Differential Equations

1181. The Laplace Equation
$$\frac{\partial^2 u}{\partial x^2} + \frac{\partial^2 u}{\partial y^2} = 0$$
applies to potential energy function $u(x,y)$ for a conservative force field in the xy-plane. Partial differential equations of this type are called *elliptic*.

1182. The Heat Equation
$$\frac{\partial^2 u}{\partial x^2} + \frac{\partial^2 u}{\partial y^2} = \frac{\partial u}{\partial t}$$

applies to the temperature distribution $u(x,y)$ in the xy-plane when heat is allowed to flow from warm areas to cool ones. The equations of this type are called parabolic.

1183. The Wave Equation
$$\frac{\partial^2 u}{\partial x^2} + \frac{\partial^2 u}{\partial y^2} = \frac{\partial^2 u}{\partial t^2}$$
applies to the displacement $u(x,y)$ of vibrating membranes and other wave functions. The equations of this type are called hyperbolic.

Chapter 11
Series

11.1 Arithmetic Series

Initial term: a_1
Nth term: a_n
Difference between successive terms: d
Number of terms in the series: n
Sum of the first n terms: S_n

1184. $a_n = a_{n-1} + d = a_{n-2} + 2d = \ldots = a_1 + (n-1)d$

1185. $a_1 + a_n = a_2 + a_{n-1} = \ldots = a_i + a_{n+1-i}$

1186. $a_i = \dfrac{a_{i-1} + a_{i+1}}{2}$

1187. $S_n = \dfrac{a_1 + a_n}{2} \cdot n = \dfrac{2a_1 + (n-1)d}{2} \cdot n$

11.2 Geometric Series

Initial term: a_1
Nth term: a_n
Common ratio: q
Number of terms in the series: n
Sum of the first n terms: S_n
Sum to infinity: S

1188. $a_n = q a_{n-1} = a_1 q^{n-1}$

1189. $a_1 \cdot a_n = a_2 \cdot a_{n-1} = \ldots = a_i \cdot a_{n+1-i}$

1190. $a_i = \sqrt{a_{i-1} \cdot a_{i+1}}$

1191. $S_n = \dfrac{a_n q - a_1}{q-1} = \dfrac{a_1 (q^n - 1)}{q-1}$

1192. $S = \lim\limits_{n \to \infty} S_n = \dfrac{a_1}{1-q}$

For $|q| < 1$, the sum S converges as $n \to \infty$.

11.3 Some Finite Series

Number of terms in the series: n

CHAPTER 11. SERIES

1193. $1 + 2 + 3 + \ldots + n = \dfrac{n(n+1)}{2}$

1194. $2 + 4 + 6 + \ldots + 2n = n(n+1)$

1195. $1 + 3 + 5 + \ldots + (2n-1) = n^2$

1196. $k + (k+1) + (k+2) + \ldots + (k+n-1) = \dfrac{n(2k+n-1)}{2}$

1197. $1^2 + 2^2 + 3^2 + \ldots + n^2 = \dfrac{n(n+1)(2n+1)}{6}$

1198. $1^3 + 2^3 + 3^3 + \ldots + n^3 = \left[\dfrac{n(n+1)}{2}\right]^2$

1199. $1^2 + 3^2 + 5^2 + \ldots + (2n-1)^2 = \dfrac{n(4n^2-1)}{3}$

1200. $1^3 + 3^3 + 5^3 + \ldots + (2n-1)^3 = n^2(2n^2-1)$

1201. $1 + \dfrac{1}{2} + \dfrac{1}{4} + \dfrac{1}{8} + \ldots + \dfrac{1}{2^n} + \ldots = 2$

1202. $\dfrac{1}{1 \cdot 2} + \dfrac{1}{2 \cdot 3} + \dfrac{1}{3 \cdot 4} + \ldots + \dfrac{1}{n(n+1)} + \ldots = 1$

1203. $1 + \dfrac{1}{1!} + \dfrac{1}{2!} + \dfrac{1}{3!} + \ldots + \dfrac{1}{(n-1)!} + \ldots = e$

11.4 Infinite Series

Sequence: $\{a_n\}$
First term: a_1
Nth term: a_n

1204. Infinite Series
$$\sum_{n=1}^{\infty} a_n = a_1 + a_2 + \ldots + a_n + \ldots$$

1205. Nth Partial Sum
$$S_n = \sum_{n=1}^{n} a_n = a_1 + a_2 + \ldots + a_n$$

1206. Convergence of Infinite Series
$$\sum_{n=1}^{\infty} a_n = L, \text{ if } \lim_{n \to \infty} S_n = L$$

1207. Nth Term Test

If the series $\sum_{n=1}^{\infty} a_n$ is convergent, then $\lim_{n \to \infty} a_n = 0$.

- If $\lim_{n \to \infty} a_n \neq 0$, then the series is divergent.

11.5 Properties of Convergent Series

Convergent Series: $\sum_{n=1}^{\infty} a_n = A$, $\sum_{n=1}^{\infty} b_n = B$
Real number: c

1208. $\sum_{n=1}^{\infty}(a_n + b_n) = \sum_{n=1}^{\infty} a_n + \sum_{n=1}^{\infty} b_n = A + B$

1209. $\sum_{n=1}^{\infty} ca_n = c\sum_{n=1}^{\infty} a_n = cA$.

11.6 Convergence Tests

1210. The Comparison Test

Let $\sum_{n=1}^{\infty} a_n$ and $\sum_{n=1}^{\infty} b_n$ be series such that $0 < a_n \leq b_n$ for all n.

- If $\sum_{n=1}^{\infty} b_n$ is convergent then $\sum_{n=1}^{\infty} a_n$ is also convergent.
- If $\sum_{n=1}^{\infty} a_n$ is divergent then $\sum_{n=1}^{\infty} b_n$ is also divergent.

1211. The Limit Comparison Test

Let $\sum_{n=1}^{\infty} a_n$ and $\sum_{n=1}^{\infty} b_n$ be series such that a_n and b_n are positive for all n.

- If $0 < \lim_{n \to \infty} \frac{a_n}{b_n} < \infty$ then $\sum_{n=1}^{\infty} a_n$ and $\sum_{n=1}^{\infty} b_n$ are either both convergent or both divergent.
- If $\lim_{n \to \infty} \frac{a_n}{b_n} = 0$ then $\sum_{n=1}^{\infty} b_n$ convergent implies that $\sum_{n=1}^{\infty} a_n$ is also convergent.

- If $\lim\limits_{n\to\infty}\dfrac{a_n}{b_n}=\infty$ then $\sum\limits_{n=1}^{\infty}b_n$ divergent implies that $\sum\limits_{n=1}^{\infty}a_n$ is also divergent.

1212. p-series

p-series $\sum\limits_{n=1}^{\infty}\dfrac{1}{n^p}$ converges for $p>1$ and diverges for $0<p\le 1$.

1213. The Integral Test

Let $f(x)$ be a function which is continuous, positive, and decreasing for all $x\ge 1$. The series

$$\sum_{n=1}^{\infty}f(n)=f(1)+f(2)+f(3)+\ldots+f(n)+\ldots$$

converges if $\int\limits_{1}^{\infty}f(x)dx$ converges, and diverges if

$$\int\limits_{1}^{n}f(x)dx\to\infty \text{ as } n\to\infty.$$

1214. The Ratio Test

Let $\sum\limits_{n=1}^{\infty}a_n$ be a series with positive terms.

- If $\lim\limits_{n\to\infty}\dfrac{a_{n+1}}{a_n}<1$ then $\sum\limits_{n=1}^{\infty}a_n$ is convergent.

- If $\lim\limits_{n\to\infty}\dfrac{a_{n+1}}{a_n}>1$ then $\sum\limits_{n=1}^{\infty}a_n$ is divergent.

- If $\lim\limits_{n\to\infty}\dfrac{a_{n+1}}{a_n}=1$ then $\sum\limits_{n=1}^{\infty}a_n$ may converge or diverge and the ratio test is inconclusive; some other tests must be used.

1215. The Root Test

Let $\sum_{n=1}^{\infty} a_n$ be a series with positive terms.

- If $\lim_{n \to \infty} \sqrt[n]{a_n} < 1$ then $\sum_{n=1}^{\infty} a_n$ is convergent.

- If $\lim_{n \to \infty} \sqrt[n]{a_n} > 1$ then $\sum_{n-1}^{\infty} a_n$ is divergent.

- If $\lim_{n \to \infty} \sqrt[n]{a_n} = 1$ then $\sum_{n=1}^{\infty} a_n$ may converge or diverge, but

no conclusion can be drawn from this test.

11.7 Alternating Series

1216. The Alternating Series Test (Leibniz's Theorem)

Let $\{a_n\}$ be a sequence of positive numbers such that
$a_{n+1} < a_n$ for all n.
$\lim_{n \to \infty} a_n = 0$.

Then the alternating series $\sum_{n=1}^{\infty} (-1)^n a_n$ and $\sum_{n=1}^{\infty} (-1)^{n-1} a_n$

both converge.

1217. Absolute Convergence

- A series $\sum_{n=1}^{\infty} a_n$ is absolutely convergent if the series $\sum_{n=1}^{\infty} |a_n|$ is convergent.

- If the series $\sum_{n=1}^{\infty} a_n$ is absolutely convergent then it is convergent.

1218. Conditional Convergence

A series $\sum_{n=1}^{\infty} a_n$ is conditionally convergent if the series is convergent but is not absolutely convergent.

11.8 Power Series

Real numbers: x, x_0

Power series: $\sum_{n=0}^{\infty} a_n x^n$, $\sum_{n=0}^{\infty} a_n (x - x_0)^n$

Whole number: n
Radius of Convergence: R

1219. Power Series in x

$$\sum_{n=0}^{\infty} a_n x^n = a_0 + a_1 x + a_2 x^2 + \ldots + a_n x^n + \ldots$$

1220. Power Series in $(x - x_0)$

$$\sum_{n=0}^{\infty} a_n (x - x_0)^n = a_0 + a_1 (x - x_0) + a_2 (x - x_0)^2 + \ldots + a_n (x - x_0)^n + .$$

1221. Interval of Convergence
The set of those values of x for which the function

$f(x) = \sum_{n=0}^{\infty} a_n (x - x_0)^n$ is convergent is called the interval of convergence.

1222. Radius of Convergence

If the interval of convergence is $(x_0 - R, x_0 + R)$ for some $R \geq 0$, the R is called the radius of convergence. It is given as

$$R = \lim_{n \to \infty} \frac{1}{\sqrt[n]{a_n}} \quad \text{or} \quad R = \lim_{n \to \infty} \left| \frac{a_n}{a_{n+1}} \right|.$$

11.9 Differentiation and Integration of Power Series

Continuous function: $f(x)$

Power series: $\sum_{n=0}^{\infty} a_n x^n$

Whole number: n

Radius of Convergence: R

1223. Differentiation of Power Series

Let $f(x) = \sum_{n=0}^{\infty} a_n x^n = a_0 + a_1 x + a_2 x^2 + \ldots$ for $|x| < R$.

Then, for $|x| < R$, $f(x)$ is continuous, the derivative $f'(x)$ exists and

$$f'(x) = \frac{d}{dx} a_0 + \frac{d}{dx} a_1 x + \frac{d}{dx} a_2 x^2 + \ldots$$

$$= a_1 + 2a_2 x + 3a_3 x^2 + \ldots = \sum_{n=1}^{\infty} n a_n x^{n-1}.$$

1224. Integration of Power Series

Let $f(x) = \sum_{n=0}^{\infty} a_n x^n = a_0 + a_1 x + a_2 x^2 + \ldots$ for $|x| < R$.

Then, for $|x| < R$, the indefinite integral $\int f(x)dx$ exists and

$$\int f(x)dx = \int a_0 dx + \int a_1 x dx + \int a_2 x^2 dx + \ldots$$

$$= a_0 x + a_1 \frac{x^2}{2} + a_2 \frac{x^3}{3} + \ldots = \sum_{n=0}^{\infty} a_n \frac{x^{n+1}}{n+1} + C.$$

11.10 Taylor and Maclaurin Series

Whole number: n
Differentiable function: $f(x)$
Remainder term: R_n

1225. Taylor Series

$$f(x) = \sum_{n=0}^{\infty} f^{(n)}(a) \frac{(x-a)^n}{n!} = f(a) + f'(a)(x-a) + \frac{f''(a)(x-a)^2}{2!} + \ldots$$

$$+ \frac{f^{(n)}(a)(x-a)^n}{n!} + R_n.$$

1226. The Remainder After n+1 Terms is given by

$$R_n = \frac{f^{(n+1)}(\xi)(x-a)^{n+1}}{(n+1)!}, \quad a < \xi < x.$$

1227. Maclaurin Series

CHAPTER 11. SERIES

$$f(x) = \sum_{n=0}^{\infty} f^{(n)}(0)\frac{x^n}{n!} = f(0) + f'(0)x + \frac{f''(0)x^2}{2!} + \ldots + \frac{f^{(n)}(0)x^n}{n!} + R_n$$

11.11 Power Series Expansions for Some Functions

Whole number: n
Real number: x

1228. $e^x = 1 + x + \dfrac{x^2}{2!} + \dfrac{x^3}{3!} + \ldots + \dfrac{x^n}{n!} + \ldots$

1229. $a^x = 1 + \dfrac{x \ln a}{1!} + \dfrac{(x \ln a)^2}{2!} + \dfrac{(x \ln a)^3}{3!} + \ldots + \dfrac{(x \ln a)^n}{n!} + \ldots$

1230. $\ln(1+x) = x - \dfrac{x^2}{2} + \dfrac{x^3}{3} - \dfrac{x^4}{4} + \ldots + \dfrac{(-1)^n x^{n+1}}{n+1} \pm \ldots,\ -1 < x \leq 1.$

1231. $\ln \dfrac{1+x}{1-x} = 2\left(x + \dfrac{x^3}{3} + \dfrac{x^5}{5} + \dfrac{x^7}{7} + \ldots\right),\ |x| < 1.$

1232. $\ln x = 2\left[\dfrac{x-1}{x+1} + \dfrac{1}{3}\left(\dfrac{x-1}{x+1}\right)^3 + \dfrac{1}{5}\left(\dfrac{x-1}{x+1}\right)^5 \ldots\right],\ x > 0.$

1233. $\cos x = 1 - \dfrac{x^2}{2!} + \dfrac{x^4}{4!} - \dfrac{x^6}{6!} + \ldots + \dfrac{(-1)^n x^{2n}}{(2n)!} \pm \ldots$

CHAPTER 11. SERIES

1234. $\sin x = x - \dfrac{x^3}{3!} + \dfrac{x^5}{5!} - \dfrac{x^7}{7!} + \ldots + \dfrac{(-1)^n x^{2n+1}}{(2n+1)!} \pm \ldots$

1235. $\tan x = x + \dfrac{x^3}{3} + \dfrac{2x^5}{15} + \dfrac{17x^7}{315} + \dfrac{62x^9}{2835} + \ldots,\ |x| < \dfrac{\pi}{2}.$

1236. $\cot x = \dfrac{1}{x} - \left(\dfrac{x}{3} + \dfrac{x^3}{45} + \dfrac{2x^5}{945} + \dfrac{2x^7}{4725} + \ldots\right),\ |x| < \pi.$

1237. $\arcsin x = x + \dfrac{x^3}{2 \cdot 3} + \dfrac{1 \cdot 3 x^5}{2 \cdot 4 \cdot 5} + \ldots + \dfrac{1 \cdot 3 \cdot 5 \ldots (2n-1) x^{2n+1}}{2 \cdot 4 \cdot 6 \ldots (2n)(2n+1)} + \ldots,$
$|x| < 1.$

1238. $\arccos x = \dfrac{\pi}{2} - \left(x + \dfrac{x^3}{2 \cdot 3} + \dfrac{1 \cdot 3 x^5}{2 \cdot 4 \cdot 5} + \ldots + \dfrac{1 \cdot 3 \cdot 5 \ldots (2n-1) x^{2n+1}}{2 \cdot 4 \cdot 6 \ldots (2n)(2n+1)} + \ldots\right),$
$|x| < 1.$

1239. $\arctan x = x - \dfrac{x^3}{3} + \dfrac{x^5}{5} - \dfrac{x^7}{7} + \ldots + \dfrac{(-1)^n x^{2n+1}}{2n+1} \pm \ldots,\ |x| \leq 1.$

1240. $\cosh x = 1 + \dfrac{x^2}{2!} + \dfrac{x^4}{4!} + \dfrac{x^6}{6!} + \ldots + \dfrac{x^{2n}}{(2n)!} + \ldots$

1241. $\sinh x = x + \dfrac{x^3}{3!} + \dfrac{x^5}{5!} + \dfrac{x^7}{7!} + \ldots + \dfrac{x^{2n+1}}{(2n+1)!} + \ldots$

11.12 Binomial Series

Whole numbers: n, m
Real number: x
Combinations: ^nC_m

1242. $(1+x)^n = 1 + {^nC_1}x + {^nC_2}x^2 + \ldots + {^nC_m}x^m + \ldots + x^n$

1243. $^nC_m = \dfrac{n(n-1)\ldots[n-(m-1)]}{m!}$, $|x|<1$.

1244. $\dfrac{1}{1+x} = 1 - x + x^2 - x^3 + \ldots$, $|x|<1$.

1245. $\dfrac{1}{1-x} = 1 + x + x^2 + x^3 + \ldots$, $|x|<1$.

1246. $\sqrt{1+x} = 1 + \dfrac{x}{2} - \dfrac{x^2}{2\cdot 4} + \dfrac{1\cdot 3 x^3}{2\cdot 4\cdot 6} - \dfrac{1\cdot 3\cdot 5 x^4}{2\cdot 4\cdot 6\cdot 8} + \ldots$, $|x|\le 1$.

1247. $\sqrt[3]{1+x} = 1 + \dfrac{x}{3} - \dfrac{1\cdot 2 x^2}{3\cdot 6} + \dfrac{1\cdot 2\cdot 5 x^3}{3\cdot 6\cdot 9} - \dfrac{1\cdot 2\cdot 5\cdot 8 x^4}{3\cdot 6\cdot 9\cdot 12} + \ldots$, $|x|\le 1$.

11.13 Fourier Series

Integrable function: $f(x)$
Fourier coefficients: a_0, a_n, b_n
Whole number: n

1248. $f(x) = \dfrac{a_0}{2} + \sum\limits_{n=1}^{\infty}(a_n \cos nx + b_n \sin nx)$

1249. $a_n = \dfrac{1}{\pi}\int\limits_{-\pi}^{\pi} f(x)\cos nx\, dx$

1250. $b_n = \dfrac{1}{\pi}\int\limits_{-\pi}^{\pi} f(x)\sin nx\, dx$

Chapter 12
Probability

12.1 Permutations and Combinations

Permutations: ^nP_m
Combinations: ^nC_m
Whole numbers: n, m

1251. Factorial
$$n! = 1 \cdot 2 \cdot 3 \ldots (n-2)(n-1)n$$
$$0! = 1$$

1252. $^nP_n = n!$

1253. $^nP_m = \dfrac{n!}{(n-m)!}$

1254. Binomial Coefficient
$$^nC_m = \binom{n}{m} = \dfrac{n!}{m!(n-m)!}$$

1255. $^nC_m = {^nC_{n-m}}$

1256. $^nC_m + {^nC_{m+1}} = {^{n+1}C_{m+1}}$

1257. $^nC_0 + {}^nC_1 + {}^nC_2 + \ldots + {}^nC_n = 2^n$

1258. Pascal's Triangle

Row 0						1						
Row 1					1		1					
Row 2				1		2		1				
Row 3			1		3		3		1			
Row 4		1		4		6		4		1		
Row 5	1		5		10		10		5		1	
Row 6	1	6		15		20		15		6		1

12.2 Probability Formulas

Events: A, B
Probability: P
Random variables: X, Y, Z
Values of random variables: x, y, z
Expected value of X: μ
Any positive real number: ε
Standard deviation: σ
Variance: σ^2
Density functions: $f(x)$, $f(t)$

1259. Probability of an Event
$$P(A) = \frac{m}{n},$$
where
m is the number of possible positive outcomes,
n is the total number of possible outcomes.

1260. Range of Probability Values
$0 \leq P(A) \leq 1$

1261. Certain Event
$P(A) = 1$

1262. Impossible Event
$P(A) = 0$

1263. Complement
$P(\overline{A}) = 1 - P(A)$

1264. Independent Events
$P(A/B) = P(A)$,
$P(B/A) = P(B)$

1265. Addition Rule for Independent Events
$P(A \cup B) = P(A) + P(B)$

1266. Multiplication Rule for Independent Events
$P(A \cap B) = P(A) \cdot P(B)$

1267. General Addition Rule
$P(A \cup B) = P(A) + P(B) - P(A \cap B)$,
where
$A \cup B$ is the union of events A and B,
$A \cap B$ is the intersection of events A and B.

1268. Conditional Probability
$P(A/B) = \dfrac{P(A \cap B)}{P(B)}$

1269. $P(A \cap B) = P(B) \cdot P(A/B) = P(A) \cdot P(B/A)$

CHAPTER 12. PROBABILITY

1270. Law of Total Probability

$$P(A) = \sum_{i=1}^{m} P(B_i) P(A/B_i),$$

where B_i is a sequence of mutually exclusive events.

1271. Bayes' Theorem

$$P(B/A) = \frac{P(A/B) \cdot P(B)}{P(A)}$$

1272. Bayes' Formula

$$P(B_i/A) = \frac{P(B_i) \cdot P(A/B_i)}{\sum_{k=1}^{m} P(B_i) \cdot P(A/B_i)},$$

where
B_i is a set of mutually exclusive events (hypotheses),
A is the final event,
$P(B_i)$ are the prior probabilities,
$P(B_i/A)$ are the posterior probabilities.

1273. Law of Large Numbers

$$P\left(\left|\frac{S_n}{n} - \mu\right| \geq \varepsilon\right) \to 0 \text{ as } n \to \infty,$$

$$P\left(\left|\frac{S_n}{n} - \mu\right| < \varepsilon\right) \to 1 \text{ as } n \to \infty,$$

where
S_n is the sum of random variables,
n is the number of possible outcomes.

1274. Chebyshev Inequality

$$P(|X - \mu| \geq \varepsilon) \leq \frac{V(X)}{\varepsilon^2},$$

where $V(X)$ is the variance of X.

1275. Normal Density Function

$$\varphi(x) = \frac{1}{\sigma\sqrt{2\pi}} e^{-\frac{(x-\mu)^2}{2\sigma^2}},$$

where x is a particular outcome.

1276. Standard Normal Density Function

$$\varphi(z) = \frac{1}{\sqrt{2\pi}} e^{-\frac{z^2}{2}}$$

Average value $\mu = 0$, deviation $\sigma = 1$.

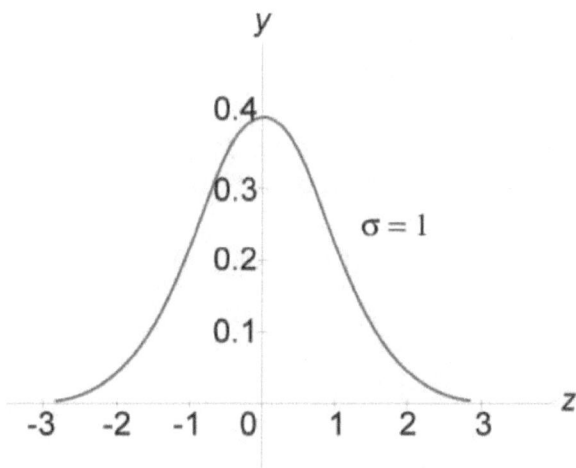

Figure 210.

1277. Standard Z Value

$$Z = \frac{X - \mu}{\sigma}$$

1278. Cumulative Normal Distribution Function

$$F(x) = \frac{1}{\sigma\sqrt{2\pi}} \int_{-\infty}^{x} e^{-\frac{(t-\mu)^2}{2\sigma^2}} dt,$$

where
x is a particular outcome,
t is a variable of integration.

1279. $P(\alpha < X < \beta) = F\left(\dfrac{\alpha - \mu}{\sigma}\right) - F\left(\dfrac{\beta - \mu}{\sigma}\right),$

where
X is normally distributed random variable,
F is cumulative normal distribution function,
$P(\alpha < X < \beta)$ is interval probability.

1280. $P(|X - \mu| < \varepsilon) = 2F\left(\dfrac{\varepsilon}{\sigma}\right),$

where
X is normally distributed random variable,
F is cumulative normal distribution function.

1281. Cumulative Distribution Function

$$F(x) = P(X < x) = \int_{-\infty}^{x} f(t)dt,$$

where t is a variable of integration.

1282. Bernoulli Trials Process
$\mu = np$, $\sigma^2 = npq$,
where
 is a sequence of experiments,
p is the probability of success of each experiments,
q is the probability of failure, $q = 1 - p$.

1283. Binomial Distribution Function

$$b(n, p, q) = \binom{n}{k} p^k q^{n-k},$$

$\mu = np$, $\sigma^2 = npq$,
$f(x) = (q + pe^x)^n$,
where
n is the number of trials of selections,
p is the probability of success,
q is the probability of failure, $q = 1 - p$.

1284. Geometric Distribution
$$P(T = j) = q^{j-1}p,$$
$$\mu = \frac{1}{p}, \quad \sigma^2 = \frac{q}{p^2},$$
where
T is the first successful event is the series,
j is the event number,
p is the probability that any one event is successful,
q is the probability of failure, $q = 1 - p$.

1285. Poisson Distribution
$$P(X = k) \approx \frac{\lambda^k}{k!} e^{-\lambda}, \quad \lambda = np,$$
$$\mu = \lambda, \quad \sigma^2 = \lambda,$$
where
λ is the rate of occurrence,
k is the number of positive outcomes.

1286. Density Function
$$P(a \leq X \leq b) = \int_a^b f(x) dx$$

1287. Continuous Uniform Density
$$f = \frac{1}{b-a}, \quad \mu = \frac{a+b}{2},$$

where f is the density function.

1288. Exponential Density Function
$$f(t) = \lambda e^{-\lambda t}, \ \mu = \lambda, \ \sigma^2 = \lambda^2$$
where t is time, λ is the failure rate.

1289. Exponential Distribution Function
$$F(t) = 1 - e^{-\lambda t},$$
where t is time, λ is the failure rate.

1290. Expected Value of Discrete Random Variables
$$\mu = E(X) = \sum_{i=1}^{n} x_i p_i,$$
where x_i is a particular outcome, p_i is its probability.

1291. Expected Value of Continuous Random Variables
$$\mu = E(X) = \int_{-\infty}^{\infty} x f(x) dx$$

1292. Properties of Expectations
$$E(X+Y) = E(X) + E(Y),$$
$$E(X-Y) = E(X) - E(Y),$$
$$E(cX) = cE(X),$$
$$E(XY) = E(X) \cdot E(Y),$$
where c is a constant.

1293. $E(X^2) = V(X) + \mu^2,$
where
$\mu = E(X)$ is the expected value,
$V(X)$ is the variance.

1294. Markov Inequality
$$P(X>k) \le \frac{E(X)}{k},$$
where k is some constant.

1295. Variance of Discrete Random Variables
$$\sigma^2 = V(X) = E[(X-\mu)^2] = \sum_{i=1}^{n}(x_i-\mu)^2 p_i,$$
where
x_i is a particular outcome,
p_i is its probability.

1296. Variance of Continuous Random Variables
$$\sigma^2 = V(X) = E[(X-\mu)^2] = \int_{-\infty}^{\infty}(x-\mu)^2 f(x)dx$$

1297. Properties of Variance
$$V(X+Y) = V(X) + V(Y),$$
$$V(X-Y) = V(X) + V(Y),$$
$$V(X+c) = V(X),$$
$$V(cX) = c^2 V(X),$$
where c is a constant.

1298. Standard Deviation
$$D(X) = \sqrt{V(X)} = \sqrt{E[(X-\mu)^2]}$$

1299. Covariance
$$cov(X,Y) = E[(X-\mu(X))(Y-\mu(Y))] = E(XY) - \mu(X)\mu(Y),$$
where
X is random variable,
$V(X)$ is the variance of X,
μ is the expected value of X or Y.

1300. Correlation

$$\rho(X,Y) = \frac{\operatorname{cov}(X,Y)}{\sqrt{V(X)V(Y)}},$$

where
$V(X)$ is the variance of X,
$V(Y)$ is the variance of Y.

www.ingramcontent.com/pod-product-compliance
Lightning Source LLC
Chambersburg PA
CBHW031608210526
45464CB00004B/1477